P9-CSB-209

Enzymes:
What the Experts Know!

Enzymes:
What the Experts Know!

Your Journey to Health and
Longevity Starts Here

Tom Bohager

One World Press
2006

Copyright: © 2006 by Tom Bohager
All rights reserved. No part of this book may be reproduced in any manner without permission of the author, except in the case of quotes used in critical articles and reviews.

Cover layout and design: Jeff Zampino / Type Monkeys
Text layout and design: Kadak Graphics

ISBN: 1-4243-0795-3

Library of Congress Cataloging-in-Publication Data

Bohager, Tom.
Enzymes : what the experts know! : your journey to health and longevity starts here / Tom Bohager.
p. cm.
Includes bibliographical references and index.
ISBN 1-4243-0795-3
1. Enzymes. I. Title.

QP601.B64 2006
615'.35--dc22

2006027336

This book was printed in the USA by
ONE WORLD PRESS
1042 Willow Creek Road
Prescott, AZ 86301
800-250-8171
production@oneworldpress.com
www.OneWorldPress.com

06 07 08 09 10   5 4 3 2

CONTENTS

PART TWO

PART ONE

Introduction

Have you ever had one of those moments in life? You know, when somebody says something profound or you read something that turns the proverbial light on for you? Perhaps it's a moment when you realize something you thought you knew, you really *didn't* know. Then suddenly out of nowhere you "get it," and the light goes on! (Actually it's more like a beacon.)

I predict that somewhere between the first and the twelfth chapters of this book, you will have a light go on for you. It will be something that will be so clear and so profound, yet so simple you won't believe it took this long to figure out. You will say to yourself, it just can't be that simple; if it were, surely more people would know or someone would have told me!

You are about to learn a simple truth—a truth that will change your life and the lives of many of your loved ones. It is the truth about what it really takes to live a long and healthy life.

Consider This!

Often an incident or a crisis in our lives will spark the acknowledgement of our inevitable mortality. It happens at different times for different people. At some point we become aware that we are getting older and we will someday die. Usually around that same time we also begin to wonder how much time we might have left. Will I be like my grandfather who lived to be 98 and never really struggled with serious disease? Will I be more like Aunt Marie who was diagnosed with cancer at 47, struggled for several years with every remedy available, yet finally died at 52?

When we finally admit that our life will end, we might take living a little more seriously; perhaps we will lose those extra 13 (or 130) pounds, stop smoking, or decide to take that long-awaited vacation. Others who reason differently might rationalize that they can't do much to alter their life expectancy, so they decide not to worry! Which type of person are you?

Consider this! Marion Higgins is 111 years old. According to the non-profit Gerontology Research Group, she is referred to as a "validated super centenarian," someone that can document his or her age over 110 years. Ms. Higgins is one of 61 "validated" members living in the world today. The current eldest member of this group was born on June 29, 1890 and recently celebrated her 115th birthday. In truth, more than 54,000 Americans are 100 years or older, and of that group, 1,388 were older than 110 years according to the 2000 U.S census. More interesting is the fact that this is the first time the U.S. census has provided numbers for the 110 year-old age group; this population that continues to grow is creating significant interest in the science of longevity. While the grand age that these people obtain seems to attract little attention, the truth is that these individuals have proven something profound: **The potential for people to live beyond a hundred years does exist!**

Why is it that some people are able to live to 100 or more years? What is it that gives them the grace to live that long? So many of these people are not just alive, but truly living with purpose and vitality; they are often vibrant, happy and socially active in their community. Although the pace at which they live is slower than when they were younger (say when they were 70 years old), they are active nonetheless and far from the bedridden image we have of a centenarian.

It seems that all of us know a vibrant, active person that looks younger than their age, not because of plastic surgery, but because they are blessed. We make the assumption that they are blessed with the genes that keep them healthy, young looking, vibrant and happy. We assume that it is not because of anything they have done personally; rather it was the luck of the draw. However, can genetics be the reason some individuals live longer than others? Not necessarily. Some researchers believe that genetics only accounts for 15% of what determines lifespan. Others put this number closer to 5%. Either way we can conclude that the role of genetics is a small one in longevity measurement. Truly, it is a combination of how we live, where we live, what we eat, how much we eat, how much we exercise, and what we do for work that actually determines how long we live. In other words, our lifestyle choices establish our lifespan. If Marion can live to be 111 years old, so can you!

Our Life Potential

What a shame it would be if we have this amazing potential and only live to be 70! In effect, we would lose 41 years of life. If you are 25, you might be thinking, "Great! I have another 86 years on this planet, provided I don't get hit by a bus or fall off a building." Accidents come to mind, but unfortunately you never think about the choices you make that shorten your potential lifespan: unhealthy food, tobacco, alcohol in excess, lack of sleep, stress, lack of exercise. The list goes on. One day you wake up and you are 60 years old and look and feel unwell. You think to yourself there is no way I am going to make it to 111! Guess what, you're right.

Your potential *was* 111, but not anymore.

We all have a life potential, but that potential varies. Our life potential dictates how long we have to live under ideal circumstances. In other words, we all have a little something different to work with. Some of us have the potential to live to be 111 like Marion Higgins, yet because of the choices we have made and the chances we have taken, we may only live to be 70.

So, now what? Well, I am happy to tell you that you have a simple choice. You can continue on the road you are currently on or get off that road and start a new journey. You can travel a new road that squeezes every second you have left out of your existing life's potential; not a life of denial where everything you put in your mouth is carefully measured and calculated. That would just take the fun out of life, no matter how long you live. No, I'm recommending a journey that is defined by a newfound knowledge of what is ultimately burning up your potential years. Here is the key: I am asking you to choose to do what you have, in the past, either ignored or were ignorant of.

Life Span vs. Health Span

The expression "life span" refers to essentially how long you have to live. I believe a more meaningful term is "health span," which is the length of your healthy, disease-free life. For most of us this starts when we are born (if we were born healthy) and continues until we reach a point when health is no longer a luxury we possess. When you get to that point in your life, when you have a hard time getting out of bed, when you have stomach pains, headaches, fatigue, frequent colds or worse, you think you've lost your health. Perhaps "health" ends when you are told you have a problem such as heart disease, cancer, HIV, or diabetes. Your life has not ended, but your health has. You spend more days being ill than being healthy and feeling good. However, when you reach the end of your health (unlike when you reach the end of your life), there IS something you can do about it. You do not have to accept the fact that what you have or how you feel is what you are forced to put up with. You have a choice: a choice to cave in or fight. You have a choice to do something pro-active and get your health (and life) back, or resolve to accept your destiny. The choice is yours!

That choice involves understanding one of the key elements of life and health...Enzymes! This book will help you to appreciate the importance of these amazing proteins. Then once you truly understand the power enzymes possess, you will have a newfound alternative to maintaining your health or overcoming a current health crisis. Prepare for the beacon to light up...your life.

Chapter One
What are Enzymes?

By definition, enzymes are catalysts, substances that cause a chemical reaction to move faster. For example, air is a catalyst for fire. You can make the fire in a fireplace burn faster by fanning it. Enzymes are the catalysts of biochemical reactions in living organisms. That includes the formation, breakdown and rearranging of molecules to provide organisms with the energy and materials needed to live and function. Without enzymes, these reactions would occur far too slowly for proper metabolism.

Think of your typical energy bar that contains vitamins, minerals and sugars; energy exists in the nutrients contained in the bar, but they must be unlocked before we can utilize them. A random chemical process could take years to break down the nutrients in the bar and release the energy. A catalyst can speed this process up quite dramatically by lowering the amount of energy it takes to start the process of releasing energy. Though enzymes are catalysts, not all catalysts are enzymes.

Enzymes are made up of amino acids. Generally speaking a combination of amino acids is called a protein. Many refer to enzymes as unique proteins that are biologically active or contain energy. It is this energy that makes it possible for enzymes to perform the work of life. However, enzymes should not be simplified by putting them in the same category as all other proteins. The activity (energy) can actually be measured in a laboratory by exposing it to a substance that the enzyme has the ability to break down (a substrate). This energy contained in enzymes is not unlimited. Rather, enzymes will work until they wear out or run out of energy. Once the enzyme no longer contains any energy, it will cease being a catalyst and is now like any other protein; it will be utilized as a protein source for the body.

Contrary to popular belief, enzymes are not alive, nor are they living cells and therefore they do not die. They either become "inactive" or they become "denatured." When inactive, the enzyme is not acting on, catalyzing or digesting anything. This may be due to a lack of water, an incompatible temperature or an incompatible pH (acid/alkaline range). Enzymes are very sensitive to their surroundings and as such, they are most active when their surroundings become more conducive to their work. However, when enzymes are exposed to temperature or pH extremes, they may become denatured. When denatured, the enzyme loses all activity and can no longer serve as a catalyst under any condition. Enzymes only work within specific temperature and pH ranges; in fact, within these ranges, they have an even narrower range in which they function most effectively. This is called the "optimal pH" and the "optimal temperature."

On the following page is a computer-generated image of an enzyme. Note that proteins are usually drawn in a way that resembles a pearl necklace. The amino acid sequences are in a strand that makes up a protein. Yet what you see depicted below resembles nothing of the sort. It is for this reason that many believe classifying enzymes as proteins is oversimplifying what they really are and how they differ from all other proteins.

It appears that an enzyme is a collapsed protein. By some miracle, a sequence of amino acids (a protein) collapses in on itself and as a result an "active site" is created. In the image below it is represented by the two balls in the middle left. It is this that separates it from all other proteins. Here is where activity or energy is focused and all of the work takes place.

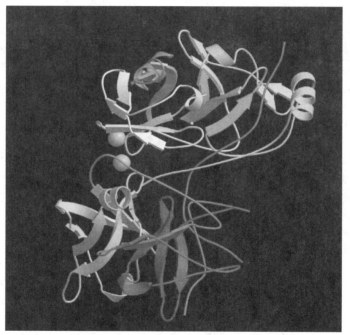

Image 1: *Computer rendition of an enzyme*

Enzyme Types

Enzymes are categorized in different ways and classified by the type of chemical reaction catalyzed. One category includes hydrolases and hydrases, the addition or removal of water. Another category includes oxidases and dehydrogenases, the transfer of electrons from one chemical to another. There are also categories that include enzymes that transfer radicals (charged electrons), enzymes that assist with the changing geometry or structure of a molecule, and enzymes that join molecules together.

However, for the sake of nutrition and enzyme therapy, all enzymes being discussed in this book are included in one of three basic categories; proteases, lipases and amylases (also known as carbohydrases). Each of these categories plays a specific role metabolically (creating energy in the body) and digestively (assisting with extracting energy from nutrients). The job of **protease** is to break down or hydrolyze proteins; **lipases** break down lipids (fats), and **amylases** break down carbohydrates. It is important to understand that within these categories there are *thousands* of enzymes.

Science has so far been able to identify and name over 5,000 enzymes that our bodies manufacture and utilize, but there may be far more, perhaps tens of thousands of different enzymes within the body. Each of these enzymes is either a form of protease, lipase or amylase. Another well-known enzyme is cellulase, the enzyme that breaks down cellulose (fiber). Though cellulase is technically an amylase by definition, it is the only enzyme our body does not have the ability to manufacture. That is why occasionally this enzyme is put into a separate category.

Proteases break down proteins
Lipases break down lipids or fats
Amylases break down carbohydrates
Cellulases (technically a form of amylase)
break down cellulose

Why are there so many enzymes? Each enzyme has a specific job and it can do nothing other than what it is designed to do. For example, a protein can be made up of thousands of amino acids, all bound together. Think of it as a long chain. A protease can alter that protein by breaking certain links or bonds in the chain, but one protease can only break the chain in very specific locations. A second protease is required to continue the job. Then a third, fourth, fifth, sixth...until the protein no longer exists; in its place we have the thousands of amino acids unattached, unbound and no longer linked. It is at this moment that the body can then benefit from these simple building blocks. This process is called digestion when it is performed in the digestive tract; the enzymes that perform this function are called digestive enzymes. When an enzyme that is produced within the body performs this task in the tissue, blood stream or anywhere other than the digestive tract, they are called metabolic enzymes.

Think of enzymes as tools, with each tool having a slightly different function. You would never use a screwdriver to drive a nail into a piece of lumber. It's the same with the different categories of enzymes. Amylases can never break down proteins; their specific function is to break down carbohydrates. However, sometimes the relationship between tools is much closer. Wrenches can look alike, but they come in different sizes. No matter how hard you try, a ½" wrench will not fit a 3/4" bolt. So, even though there are similarities within the categories, (such as the numerous proteases our bodies produce) they all have a slightly different purpose; they fit a different sized bolt, figuratively speaking. Large quantities of enzymes must then be produced to cover the many specific jobs that they are responsible for.

Digestive Enzymes

The digestive process would not be possible without these

catalysts with biological activity. In the digestive system this biological activity or "energy" is what enables the enzymes to break down or digest proteins, fats and carbohydrates into their simplest components (amino acids, essential fats and sugars). Enzymes also assist in the extraction of vitamins and minerals. Then the beneficial components are delivered to the trillions of cells throughout the body, while those that are not essential or perhaps toxic are escorted out of the body. Since digestive enzymes are responsible for this process, it is safe to say that without them all of us would die of malnutrition. To better understand this process let's look a little deeper into digestion itself.

THE DIGESTIVE PROCESS

Digestion begins in the mouth where the act of chewing breaks down and grinds the food into smaller pieces to be swallowed. Think of the mouth as a food processor where mixing and grinding takes place. Here is where three different types of amylase are secreted to digest the carbohydrates we eat. The amylase is mixed with the food by the act of chewing, so it can immediately begin working on the digestion process. When we swallow the food, it travels down the esophagus to the stomach.

THE STOMACH

There are two portions of the stomach. The cardiac (upper portion) is the first stop for the food we swallow. Though nothing is secreted in this portion of the stomach, this is where the majority of the carbohydrate digestion occurs due to the activity of the amylase found in the saliva that was mixed with the food. This portion is often called the food enzyme section of the stomach since other than the small amount of carbohydrate digestion that occurs here, only the foods we eat that have not been cooked or processed are partially digested in this part of the stomach. Raw foods contain the enzymes that nature provides to break down the proteins, fats and carbohydrates contained in that particular food. When we cook and process food, we denature the enzymes naturally occurring in the living organism; then they are no longer of assistance in the digestive process.

After about 30 to 45 minutes in the cardiac portion, the food enters the pyloric (lower) section of the stomach. This is where pepsin and hydrochloric acid (HCL) are secreted for protein digestion. The amount of protein consumed and the efficiency of the individual's digestive system will dictate the length of time the food remains in this portion of the stomach (typically about two hours). The combined activity of hydrochloric acid, pepsin and muscular movement result in a thoroughly mixed, watery solution called chyme, which then leaves the stomach through the pyloric sphincter and enters the small intestine.

THE INTESTINES

When the chyme enters the duodenum (a section of the small intestine), it is mixed with additional enzymes. It is here that lipase secreted from the pancreas, and bile secreted from the gallbladder assist with fat digestion. Acid from the stomach is neutralized with bicarbonate ions, which the pancreas also manufactures and secretes. It is at this point that the body takes a type of inventory of what we have eaten and what has been digested so far. Based on this information, it then determines the additional amount of enzymes that will be needed to finish the process. This is known as the law of adaptive secretion, in which the body only manufactures the amount of enzymes needed to process the carbohydrates, protein, and fats that have reached this point undigested. The pancreas will then continue to manufacture and secrete the enzymes as needed.

As the food continues to travel through the intestinal tract, it finally enters the large intestine. This organ is basically responsible for absorbing water and electrolytes, which are actually ionic compounds. Much of this process is the body reabsorbing what it has provided in the way of gastric juices. The large intestine is also home to a number of different kinds of bacteria. These beneficial bacteria live off of some of the foods (primarily fibrous foods) that have made it this far without being digested. As a byproduct of their metabolism, they make some of the vitamins we need and limit the amount of bad bacteria that can cause disease. Finally, the waste is eliminated. The entire digestive process normally takes about three days.

Food Enzymes

The cells in plants are similar to those in both humans and animals; they produce and need enzymes to survive. When a fruit or vegetable is harvested, it is removed from its life source. When we pick an apple off the tree in our back yard, we have removed it from the branch, disconnecting it from what keeps it "alive." It is in that moment that the enzymes contained in the apple assist us in using the nutrients in the apple. If the apple goes straight from the tree to our mouth, we can take full confidence in the fact that our digestive system will be aided by the naturally occurring enzymes found in that apple. The enzymes once instrumental in cellular biochemical reactions necessary for the growth of the apple will now be contributing to the digestion of the apple in the cardiac portion of our stomach.

I'm sure you've observed an apple going "bad" because it sat in the kitchen fruit bowl too long. You didn't eat it, so it ate itself. It deteriorates in front of your eyes. The enzymes within that apple have become active in a digestive manner and the result is a spoiled apple. If you want to speed that process up, simply damage the apple in any way. The soft spots on apples are damaged areas where enzymes are particularly active (as they have been released from within the apple's cells) and the apple is now being digested. When we chew the apple, we are literally speeding up the reaction we have watched in our kitchen fruit bowl. Add to this the 98-degree temperature inside our mouths, the naturally occurring amylases in our saliva, and the water contained in the apple, and you now have an environment perfectly suited to induce the greatest food enzyme activity possible. The apple is being digested primarily by the enzymes that at one time were active in keeping the apple alive. They have become digestive enzymes and have taken the burden of digestion off our digestive system. There is little doubt that enzymes will be secreted to assist in the delivery of the nutrients of the apple to our cells, but far fewer are needed than if the apple had been cooked or processed. If that same apple were used in an apple pie that were baked or in a stewed apple dish, the enzymes would be denatured and of no use to us digestively. Upon eating cooked apples, the digestive system will be called on to produce the enzymes needed to digest the cooked food.

> **Food enzymes are introduced to the body through the raw foods we eat. However, raw food manifests only enough enzymes to digest that particular food, not enough to be stored in the body for later use. The cooking and processing of food destroys all of their enzymes.**

Metabolic Enzymes

Though you have probably heard of enzymes and you might have known a little about the essential aspects of digestive enzymes, very few people know about or can accurately describe the importance of metabolic enzymes. Metabolic enzymes are defined as any enzyme produced within the body that is not utilized for digestion. They have been called the spark of life, the energy of life, and the vitality of life because without these enzymes we would not be able to hear, feel, think, walk, talk, breathe or live. They are the enzymes that make biochemical reactions possible within the cells for detoxification and energy production.

Every organ, every tissue, and all 100 trillion cells in our body depend upon the reaction of metabolic enzymes and their energy. All living cells produce metabolic enzymes, although the pancreas, liver and gallbladder play a vital role in determining the amount of metabolic enzymes the body is capable of producing. With such an important role in the body, you may wonder why it is that few of us have ever heard of metabolic enzymes. The reason is that enzymes that are responsible for every function of the body cannot be produced in a lab, encapsulated, bottled and sold to the consumer. If they could, the metabolic enzyme pills would be in every health food store and pharmacy in the world and would be responsible for curing nearly every disease.

Knowing that fact is really just the tip of the proverbial iceberg! Enzymes are responsible for the digestion of food, the assimilation of the nutrients found in food, the elimination of the non-essential and toxic ingredients, and with rare exception, every single reaction that takes place in every living cell. Without enzymes, life would be impossible. Enzymes are life! But it doesn't end there; a lack of enzymes means death! Or to put it another way, a lack of enzymes is the cause of most disease, which leads to death!

Has the light gone on for you yet? Here is the simplest of truths; perhaps so simple you may have overlooked it. **If you want to stay healthy, you need to support the body's mechanism for producing and conserving enzyme production.** *Most diseases known to man are the result of an enzyme imbalance.* In fact, it can be stated that any disease that is not directly linked to a nutritional deficiency in which the person's digestive system works as it should, is due to an enzyme imbalance.

If nutritional needs have been met and a person is sick, an enzyme imbalance or deficiency is to blame. Often this is a combination of lack of enzymes that lead to a disease state; sometimes it is just the lack of one particular enzyme.

Arthur Kornberg won a Nobel Prize in medicine for his study of enzymes. In his book, *For the Love of Enzymes: The Odyssey of a Biochemist* he writes, "By 1982 some 1,400 diseases, each due to a defect in a single gene, had been described in medical literature...For most of the diseases, the biochemical basis is still unknown. For approximately 200 of them, the disease is known to be due to a deficiency or malnutrition of a single enzyme."

Let's consider what it means that *nearly every disease known to man is the result of an enzyme imbalance.* If your nutritional needs have been met and you have a cold, acid reflux, cancer, AIDS, hormonal imbalances, eczema, arthritis, chronic fatigue, or flu, then you have an enzyme imbalance. In nearly every case the imbalance is a deficiency in the appropriate enzyme or type of enzymes needed to overcome the issue. What if you could take an enzyme pill to overcome the deficiency? Is there an enzyme pill for a cold, acid reflux, cancer, AIDS, hormonal imbalances, eczema, arthritis, chronic fatigue or the flu? Not exactly....but there is hope.

*Dr. Edward Howell in his book *Enzyme Nutrition* makes a case that it is unfair to categorize enzymes as proteins. He suggests that to call an enzyme a protein is to oversimplify what an enzyme really is and does; no other "protein" known has the activity or energy needed to speed up biochemical reactions. Thus he felt enzymes should not be simplified by putting them in the same category as all other proteins. [*Enzyme Nutrition*. NY: Avery, Penguin Putnam, 1985, p. 5.]

Chapter Two
The History of Enzyme Therapy

The history of enzyme therapy is fascinating. At times it is similar to a classic mystery–small clues over many years reveal the true identity and functionality of enzymes. Though enzyme therapy has been around for centuries, people did not know that enzymes were the reason it worked. Two examples of this are the South American Indians and the ancient Israelites. In South America for example, papaya leaf was used for its ability to support digestion and promote healing. Papaya is the source of an enzyme known as papain; it is still widely used today for these same purposes. Moreover, the Bible (2 Kings 20:7) recommends the use of figs for boils to the Israelite nation. The fig contains an enzyme, ficin, which is still used today in some skin gels and dietary supplements. In more modern times, enzymes have been the subject of studies for digestive issues, cancer, immune support, inflammatory issues and many other debilitating diseases. If we take a look back to the late 1800s and into the first part of the 20th century, we can appreciate why enzyme therapy holds such promise.

In the Far East, an age-old tradition was common where molds (fungus) called koji were used in the production of certain foods and flavoring additives based on the soy plant. Examples of this are shoyu (soy sauce) and miso, both of which are made from soybeans and originated in Japan. During the fermentation process, miso develops a complex and distinct flavor. Additionally, natto, produced from the fermentation of soy beans by means of a bacterium called *Bacillus subtillus*, has been eaten by the Japanese for hundreds of years. It has been credited with many medicinal properties and only recently have scientists given the enzyme found in natto a name: nattokinase.[1,2,3,4]

In 1891 Dr. Jokichi Takamine filed patent applications for "Taka koji" from *Aspergillus oryzae* (fungi that is rich with enzyme activity). This formed the basis for Dr. Takamine's fermentation process for the industrial production of a fungal amylase, the first of its kind. The method of fermentation suggested by Takamine is still used in the production of certain enzymes today. In 1894 Takamine moved his family to the U.S. and opened his own research laboratory in New York City. Takamine allowed the pharmaceutical company Parke, Davis & Company to produce his enzyme, takadiastase, on a commercial scale; it is still in use today as digestive aid.[5,6]

Later in 1926, Dr. James B. Sumner was able to determine that enzymes are actually proteins. As a result of this, he successfully crystallized an enzyme marking the beginning of commercial enzyme production and their use as dietary supplements and food additives.[7]

Elsewhere, enzyme therapy was heading in a slightly different direction through the investigative use of enzymes from animals. By 1930 there were two directions that enzyme therapy was moving in. One involved the study of fungal enzymes and the other focused on the study of animal enzymes.

The following section is an attempt to credit the individuals who have been responsible for this research throughout the years. It is impossible to mention all of them, but here are some of the key players and what their contribution was.

The Founding Fathers (and Mothers)
Dr. John Beard

In the early 1900s John Beard was an embryologist living in Scotland. His main research interest was the placenta. Beard observed that the cells that eventually become the placenta behaved like cancer cells. He also noted that the placenta stops growing on day 56 of the human pregnancy, which interestingly is the same day the fetus' pancreas begins to function. He came to the conclusion that the fetus' pancreas secreted something that stopped the growth of the placenta and hypothesized that the same substance might stop the growth of malignant cancer.

Beard conducted experiments with the juices extracted from young animal pancreases to test his theory. The juices were injected into cancer tumors and the tumors shrank in both animals and humans. Beard's work was published in the *Journal of the American Medical Association*. In 1911, Dr. Beard published a monograph entitled "The Enzyme Therapy of Cancer," summarizing his therapy and the supporting evidence.[8] After Dr. Beard's death in 1923, the enzyme therapy was largely forgotten.

From time to time some alternative therapists have "rediscovered" Dr. Beard's work and used pancreatic proteolytic enzymes as a treatment for cancer.[9] Today this same therapy is being researched by **Dr. Nicholas Gonzalez, M.D.** who has published several studies on the effects of pancreatic enzymes on individuals diagnosed with cancer. You can read about his work at www.dr-gonzalez.com.

Dr. Max Wolf

Shortly after finishing medical school, Dr. Wolf was appointed professor of medicine at Fordham University in New York. There he became aware of the key role enzymes play in the vital process of life itself. He was one of the first to speculate about the therapeutic possibilities of enzymes.

He was able to convince **Dr. Helen Benitez** to join him from her post in the neurosurgical department at Columbia University. After conducting hundreds of tests, they concluded that enzymes were the missing factor in people that developed cancer. They also discovered that enzymes have an anti-inflammatory effect. Along with **Dr. Karl Ransberger**, they began to isolate many of the dozens of known enzymes that were responsible for anti-inflammatory activities. After years of testing, they created an enzyme formula, Wobenzym, (derived from a combination of their names, Wolf and Benitez). Wobenzym is comprised of the animal-based enzymes called pancreatin, trypsin and chymotrypsin, the plant enzymes bromelain and papain, and the bioflavonoid rutin.

Dr. Francis Pottenger

From 1932 to 1942, Dr. Francis Pottenger began a study that ran for ten years, covering four generations of over 900 cats. In this study, Dr. Pottenger controlled the type of food the cats were fed. One group was fed only raw food and unpasteurized milk, while the others were fed a combination of cooked and processed foods. Dr. Pottenger then recorded his observations with exacting measurements and photographs.

The cats fed cooked and pasteurized milk developed common degenerative diseases such a diabetes and arthritis, while the group of cats fed *only raw food* prospered, living much longer than the cats from the other groups. Dr. Pottenger reported the underlying nutritional factor had to be a substance that was destroyed by the heat used in the cooking and pasteurization processes; the raw foods not exposed to this processing maintained this substance (enzymes), while the cooked and processed food did not.[10,11]

Dr. Edward Howell

At the same time that Dr. Pottenger was overseeing the clinical study in California and Dr. Wolf was researching in New York, Dr. Edward Howell of Chicago was questioning the use of cooked and processed food for human consumption. He found that heating food to 118°F for more than 15 minutes destroyed all of the enzymes that naturally occur in raw food. In 1940, Dr. Howell began to investigate whether or not chronic degenerative disease was a matter of a severe enzyme deficiency.

Dr. Howell wrote two books reporting his life's work: *Enzyme Nutrition* as well as *Food Enzymes for Health and Longevity*. Some of the most important and profound discoveries about enzymes and enzyme therapy are contained in the pages of these two books. Here are a few examples:

1) Mammals have a pre-digestive stomach; he called it the "food enzyme stomach." In humans, it is the uppermost portion of the stomach. It is here that enzymes found in raw food pre-digest what has been consumed.

2) When cooked food is eaten a constant drain of enzymes from the immune system and other important organs is observed.

3) Any organ or gland will grow more cells, becoming larger because the demand placed on it exceeds its ability to function, much the same as a person's muscles will grow if they lift weights. He found that, in particular, human pancreases are 2 to 3 times heavier and larger in proportion to body weight as compared to the pancreases of other mammals. He attributed this to consumption of an excessive amount of cooked foods which place a large demand on the pancreas for digestive enzyme production.[12,13]

In the 1940s Dr. Howell founded the National Enzyme Company (NEC). The company today is one of the largest buyers of imported enzymes and manufactures finished goods for nutritional companies. While doctors Beard and Wolf used animal-based enzymes produced from the pancreas of animals, Dr. Howell used certain species of fungus to grow highly concentrated plant-based enzymes.

There are two schools of thought regarding the source of enzymes used in enzyme therapy. They are divided on whether the animal or plant-based source is more effective. The next chapter discusses the differences in detail.

Additional Contributors
Gabriel Cousens, M.D., N.D.

Dr. Cousens runs the Tree of Life Rejuvenation Center in the U.S. and has written several books. My favorite is *Conscious Eating* in which he talks about the importance of raw food, enzymes, enzyme therapy and pH balance. In my opinion, this is one of the most thorough and balanced approaches to nutrition, health and healing ever written. It is based on his experience and success in the clinic he founded and thus has real practical benefits.

Ellen Cutler, M.D.

Dr. Cutler, the "Enzyme Empress," as she is called by her patients, is the bestselling author of four books, an internationally recognized teacher and an eminent public speaker. Dr. Cutler dedicates herself to further scientific research, teaching, writing, and the ongoing consultation of clients with particularly complex conditions from around the world. Her latest book *Enzymes, MicroMiracles* is a culmination of her life's work.

During the past 25 years in clinical practice and subsequent to successfully healing her own life-threatening condition with enzymes, Dr. Cutler found that food sensitivities and improper digestion contribute to a surprisingly extensive number of ailments. These include obesity, skin problems, Chronic Fatigue Syndrome, immune disorders, asthma, some forms of cancer and of course digestive diseases such as colitis. Dr. Cutler's treatment approach incorporates her profound knowledge of both conventional and natural healing.

Karen DeFelice, M.S.

Karen has a Master's Degree in Science and has written *two books, Enzymes for Autism and other Neurological Conditions* and *Enzymes for Digestive Health and Nutritional Wealth.* She and her two sons deal with pervasive neurological and sensory integration dysfunctions, and have seen dramatic improvement in their conditions through the use of enzyme therapy. She has one of the most informative websites on enzymes (www.enzymestuff.com) and has never received any payment from any company for her endorsement. She speaks frequently on the topic of enzymes and autism and is an expert in this field.

Harvey Diamond

Diamond's book *Fit for Life* is still the number one nutritional book ever sold, with 12 million copies printed in 33 languages. He is known as one of the greatest proponents of raw food and proper food combining. Harvey has shown through his personal life and his writings that one of the most effective ways to support optimal health and to overcome disease is to eat a diet predominated by raw food. In his most recent book, *Living Without Pain* he describes how enzymes play a crucial role in overcoming some of the most common diseases that cause debilitating pain.

DiQuie Fuller, Ph.D.

Dr. Fullers' interest in holistic health began over 20 years ago when her daughter became ill and was given only a short time to live. She discovered her daughter's illness was as a result of her inability to properly digest, utilize, and eliminate food. With this new knowledge, her research turned to Dr. Edward Howell's work on plant enzymes. Her daughter began plant-based enzyme supplementation and experienced rapid and lasting results. From that point on, Dr. Fuller began an alternative health clinic and a line of enzyme products used exclusively by health professionals, Transformation Enzyme Corporation.

Dr. Fuller was instrumental in my education and I have nothing but the greatest respect for her and the company she founded. "Transformation" continues to educate health care professionals on the benefits of enzymes and enzyme therapy. Her book, *The Healing Power of Enzymes* describes common enzyme deficiencies, body typing and practical dietary recommendations to support healing.

Max Gerson, M.D.

Back in the 1930s and 40's, Dr. Max Gerson was treating cancer and tuberculosis at his clinic in Germany. He was among the first to discover the importance of organically grown, whole foods as he conducted research testing the reaction of all types of foods on the body's various systems. He found that raw, ripe fruits, vegetables and juices had the most healthful effects. Dr. Gerson believed that 80% of all disease could be eradicated by eliminating canned, frozen and processed foods from the American diet, foods completely devoid of enzymes. The Gerson Therapy uses intensive detoxification to eliminate wastes, regenerate the liver, reactivate the immune system and restore the body's essential defenses including the enzyme, mineral and hormone systems.

Ralph E. Holsworth, Jr., D.O., B.Sc.

Dr. Holsworth has authored several publications related to the antioxidant properties of electrolyzed water and nattokinase, a fibrinolytic enzyme. He is member of the Editorial Board for the *Journal of Applied Clinical Thrombosis and Homeostasis* and the Japan Nattokinase Research Association.

Dr. Holsworth has had the unique experience of having traveled to Germany and Japan in his pursuit of understanding enzymes in their clinical application. In Germany, Dr. Holsworth met with the late Dr. Karl Ransberger, one of the original researchers with Max Wolf in determining the benefits of animal enzymes. In Japan, he met Dr. Hiroyuki Sumi who discovered the nattokinase enzyme. Dr. Holsworth applies his knowledge and experience to support other doctors to understand the benefits of enzymes in a clinical setting.

William Kelley, D.D.S., M.S.

In 1963, William Kelley, a dentist, was diagnosed with pancreatic cancer. He rediscovered the connection between pancreatic enzymes and cancer remission (Dr. Beard's theory), and was able to cure his own cancer. He also subsequently treated hundreds of others with cancer. Kelley emphasized metabolic individuality, or metabolic typing. This theory held that no single therapy, diet, or supplement is perfect for everyone, because each person's biochemistry is totally unique and different. Kelley identified enzymes as critically important and researched the influence of genetics on the autonomic nervous system. The goal of Kelley's therapy, which included a specific diet, nutritional supplements, and detoxification, was to achieve ideal metabolic balance that would enable the body to heal itself.[12]

Roy Walford, M.D.

Dr. Walford, a professor of pathology at UCLA, is at the forefront of the research regarding calorie restriction. Calorie restriction (CR) is the theory that by limiting calories to a minimum, the aging process is slowed. The CR diet is one of the most extensively studied diets in the world. In one study, mice were divided into two groups; half were kept on a normal diet and half on a diet restricted in calories but adequate in everything else.[13] The maximum life span of the mice on the normal diet was 41 months, which would be equivalent to the maximum life span of humans of about 110 years. However, for the calorie-restricted mice, their maximum life span was pushed to 56 months, for a human equivalent of 150 years!

Similar studies on mice, rats, fish and other species have been done in numerous university laboratories during the past 50 years, and they all agree: a CR diet increases the maximum life span characteristic of the species. It also increases the population's average life span, so the two together translate into longer and healthier life.

It is important to note that the reduction of calories is not the only criteria. The calories consumed must be "nutrient dense" healthy calories.[14,15] Though I have never read or heard that Dr. Walford credits the lack of stress on the digestive system or the decreased amount of digestive enzymes required in a calorie restricted diet as being the mechanism for success, he has said "CR increases DNA repair, definitely decreases oxidative damage, and probably increases the body's own antioxidant defense systems...Dr. Richard Weindruch and I postulated some years ago that the mechanism is related to an increase in metabolic efficiency."[16] This reminds me of the role metabolic enzymes play in maintaining health within the body.

Now it's time to get into the real meat of enzyme therapy. The following chapters will highlight the latest research, both studied and hypothesized on this exciting approach to healing and health. The first topic we will address are the differences between plant-based and animal-based enzyme therapy.

References:

1. "Natto-Traditional Japanese Fermented Soy Beans with Recently Discovered Health Benefits and Novel Industrial Applications," *Enzyme Wave*, Volume 3, June 2002, Amano Enzyme, Inc., pp. 2-4.

2. "Prevent Heart Attack and Stroke with Potent Enzyme that Dissolves Deadly Blood Clots in Hours." Health Sciences Institute, March 2002.

3. Maruyama M, Sumi H., "Effect of Natto Diet on Blood Pressure." *JTTAS*, 1995.

4. Sumi H, Hamada H, Nakanishi K, Hiratani H., "Enhancement of the fibrinolytic activity in plasma by oral administration of nattokinase." *Acta Haematol* 1990; 84(3):139-43.

Enzymes: What the Experts Know!

5. Higasi, K., Structural Chemistry. In Livermore, Arthur H., *Science in Japan*; The American Association for the Advancement of Science: Washington, DC, 1965; pp. 239-266.

6. 1997, *Encyclopedia Britannica*: Takamini.

7. 1997, *Encyclopedia Britannica*: Sumner.

8. Beard, J., "The Action of Trypsin upon the Living Cells of Jensen's Mouse Tumor." *Br Med J* 4, 140-141, 1906.

9. Beard, J., *The Enzyme Treatment of Cancer*. London: Chatto and Windus, 1911.

10. Pottenger, Francis, Jr., "The Effect of Heat-Processed Foods and Metabolized Vitamin D Milk on the Dentofacial Structures of Experimental Animals," *American Journal of Orthodontics and Oral Surgery*, St Louis, MO, Vol. 32, No. 8, pp. 467-485, August 1946.

11. Pottenger, Francis, Jr., *Pottenger's Cats: A Study in Nutrition*, Price-Pottenger Foundation, Inc., La Mesa, CA, 1995.

12. Kelley, William Donald, *Cancer: Curing the Incurable Without Surgery, Chemotherapy, or Radiation*. CA: Bonita, New Century Promotions, 2001.

13 Weindruch R, Walford RL, Fligiel S, Guthrie D, "The retardation of aging in mice by dietary restriction: longevity, cancer, immunity and lifetime energy intake." *J Nutr*. 1986 Apr;116(4):641-54.

14 Walford, R.L. and Crew, M., "How dietary restriction retards aging: an integrative hypothesis," *Growth Dev Aging*, Winter 1989, 53(4) 139-140.

15. Walford, R.L., "The clinical promise of dietary restriction," *Geriatrics* 1990 Apr; 45(4):81-3, 86-7.

16. Walford, R.L., Calorie Restriction: Eat Less, Eat Better, Live Longer. *Life Extension Magazine*, Life Extension Foundation, February 1998.

Chapter Three
Animal vs. Plant-Based Enzyme Therapy

As was mentioned in the previous chapter, there are really two schools of thought when it comes to enzyme therapy. The first school hypothesizes that enzymes derived from animals are superior in overcoming a health crisis. The second school proclaims that fungal or plant-based enzymes have the advantage. In addition, there is another school, the one that states it is ALL hogwash! There are a number of "reputable" sources that say enzymes taken orally cannot survive the acid environment of the stomach. I can think of one very popular Harvard trained M. D., who believes enzymes are a waste of money for this very reason. So, before I talk about animal vs. plant-based, let's settle this notion of enzymes not surviving stomach acid once and for all.

Enzymes Can Survive Stomach Acid
It is true that enzymes are affected by pH (the relative degree of acidity versus alkalinity in an environment). It is also true that under extreme levels, whether acid or alkaline, enzymes can be denatured in this manner. However, keep in mind that within the acidic environment of the stomach

there is an enzyme known as pepsin that breaks down protein and thrives in that kind of environment.

Does the acid in the stomach measured between 2 and 3 on the pH scale (very acidic) denature or destroy supplemental enzymes? The answer has to do with which source of enzymes we are considering. Animal-sourced enzymes (trypsin, chymotrypsin and pancreatin) come from the stomach, small intestine, and pancreas of primarily cows and pigs. Since these animal enzymes are made from animal protein and become active in an alkaline pH, they are particularly sensitive to acid. Technically speaking one could say that this type of enzyme is destroyed in stomach acid. But wait, can they be protected in order to survive the environment? Yes, they can. All reputable manufacturers who sell this form of supplement "enterically coat" the enzymes. Enteric coating can be thought of as an armor that allows the enzyme to pass safely through the stomach unscathed and enter the small intestine where they can begin to do their work in an environment perfectly suited for them. Are enzymes, then, a waste of money? I hardly think so.

What about the argument that asks if the *plant*-based sources are denatured or destroyed in the acid environment? Again, this reasoning is flawed. What the critics will often say is that since enzymes are technically proteins, the body will digest them just as it would any protein. Herein lies the difficulty; while enzymes are technically protein, they are not innate, dead protein! They contain energy that catalyzes biochemical reactions. Enzymes are active in a specific pH range and are most effective in their optimal pH.

One of the truly beneficial qualities of plant-based enzymes is that most of them are active in the acid pH found in the stomach. In other words, they are actually contributing to digestion while in this acidic environment. Yet what about the enzymes that are active in a more alkaline pH? Do they become denatured in acid as well? No, they become *inactive* in acid.

You may recall from Chapter One that enzymes will become either inactive or denatured when they are in an environment that is not conducive to them. The difference between these terms is that an *inactive* enzyme will become active again when the environment changes, whereas *denatured* enzymes are those that have been destroyed and are now proteins with no energy or biological activity. It turns out that the vast majority of plant-based enzymes, especially those used in nutritional supplements are not denatured in stomach acid; they are simply temporarily inactive until they reach and environment with a more suitable pH.

The National Enzyme Company (founded by Dr. Edward Howell) set out to prove once and for all that plant based enzymes survive the acid stomach. They conducted an experiment described on the web site www.enzymeuniversity.com. I will summarize their study here.

Analyzing digestion of the same meal, the researchers carried out four different tests: the meal without the digestive enzyme blend under perfect digestive conditions, the meal with the digestive enzyme blend under perfect digestive conditions, the meal without the digestive enzyme blend with 70% reduced gastric and intestinal secretion, and the meal with the digestive enzymes under the impaired digestion model.

Samples collected at various times during the digestion process were analyzed for glucose and nitrogen content, demonstrating carbohydrate and protein digestion, respectively. The enzymes improved the bioavailability of both proteins and carbohydrates in the lumen of the small intestine, not only under impaired digestive conditions, but also in healthy human digestion.

Glucose availability was increased four-fold in the "perfect" digestive system and by seven times in the impaired digestion model. Proteins were twice as bioavailable in the impaired digestive process. The activity of any digestive enzyme supplement in the small intestine "presupposes that the enzymes survive the acidity of the stomach."

Consider this: There are many prescription drugs that contain enzymes that have been proven effective and approved for use as drugs by the Food and Drug Administration. Are the critics of this theory suggesting that the drug-enzymes are **not** affected by the acid in the stomach, while the nutritional supplements **are** destroyed? While I am aware that there are enzymes that are sensitive to acid and can be denatured in the acid part of the stomach (pyloric section), it is safe to say the vast majority are not. Yes, enzymes survive the gut! Now let's move on...

Animal-Based Enzyme Therapy

As was discussed in the previous chapter, animal-source enzymes have been studied since the late 19th century. With people like Dr. Max Wolf and more recently Dr. Nicholas Gonzalez, there is great interest in this form of therapy. The theory started with John Beard in Scotland with his observations on the placenta and the similarities to cancer cells. The premise is that the enzymes pancreatin, trypsin and chymotrypsin from animal sources contain enzymes with proteolytic activity; they have the ability to break down protein.

However, it is important to note that cancer cells are surrounded by a protein that protects it by disguising it from the immune system. These enzymes break down that protein and digest the cancer or expose the cancer to the immune system for removal. These enzymes may have a similar effect on viruses and bacteria that cause sickness since bacteria are made up of protein and viruses are protected by them. The other areas of research include inflammation, heart disease and stroke. The protein, fibrin, that contributes to these conditions is the same one that causes blood clots. By removing the excess fibrin, inflammation is reduced and the risks of inappropriate blood clots are also reduced, which greatly limits the risk of heart disease or stroke.

By some estimates over $50,000,000 has been spent on the research of these animal-source enzyme products. The evidence in the form of studies, grants, patents, pharmaceutical drugs, books and credible researchers is indisputable. There is merit to this approach and anyone who says otherwise is ignoring the evidence.

SOME FLAWS

Though there are obvious benefits of the animal enzyme approach, there are also some flaws. First, people who are vegan or vegetarian will not want to take this form of supplement since it originates with animals that must be slaughtered to extract the raw material. Second, slaughterhouse animals often come from questionable sources with both questionable health and diets. To consume this ingredient is to consume a portion (albeit small) of the animal. With the use of steroids, antibiotics and the like prevalent in raising farm animals today, many choose to stay away from this source. Third, enzyme potency is determined by how much protein, fats or carbohydrates the enzyme can break down.

The fact is that animal-based enzymes are relatively weak when compared to the plant-based sources. In fact, when testing animal enzymes against plant-based enzymes using the same substrate (digestible material), pH and assay, the plant-based enzymes break down between 10 and 100 times more protein per milligram than the animal source enzymes. So if we are going to stick with the theory above about pro-teolytic activity, it stands to reason that the higher the pro-teolytic activity, the more effective the product. Fourth, the amount of tablets consumed range from 30 to 90 a day in cases of severe distress. For most people this does not seem practical, especially since they are most likely taking other supplements and medications as well.

THE UNKNOWN

One must acknowledge that there are likely unknown factors associated with the success of this research. Is it just the proteolytic enzyme activity responsible for the success of this therapy, or is there something more? Likely, there is something more going on, something that has yet to be discovered. Because of this, it is often recommended that individuals (who are not vegetarian or vegan) diagnosed with cancer take both the plant-based and animal enzymes.

Something else needs to be said about this source of enzymes however. It has been my experience that when dealing with an individual who has extreme organ or gland deficiency (pancreatitis, pancreatic cancer, liver cancer, gall bladder failure or the like), animal source enzymes seem to fortify the glands and organs in ways that the plant-based enzymes cannot. It is what we call the "law of similar," which is the basis of homeopathy. This theory states that even though the source is animal and not human, the body recognizes it as similar. Because of this, when animal source enzymes are ingested, the body is able to recognize from which gland or organ the animal source enzyme originated; thus the body is able to use this source to fortify glands and organs in ways straight plant-based enzymes can't. This is why a thyroid glandular is recommended for individuals with a weak thyroid.

This makes sense to me and I have seen great results. When I counsel someone with these specific issues and they are able to take an animal source product, I recommend "Wobenzym N" by Naturally Vitamins or "MegaZyme" by Enzymatic Therapy, both of which are very reputable companies. Other sources that are recommended include the Michaels and Jarrow Formulas.

Plant-Based Enzyme Therapy

Plant-based enzymes currently represent about 80% of all of the enzymes sold in health food stores. The other 20% is made up of combination plant and animal enzyme products. Plant-based enzymes originate from a fungus called aspergillus. As was mentioned in the previous chapter, this fungus produces a large amount of enzymes including proteases, lipases, amylases and cellulases, naturally active in a broad pH range. As the most potent plant source, the enzymes are often called fungal or microbial enzymes. However, most people in the nutrition industry refer to them as plant-based. This is appropriate since the aspergillus is grown in a lab on different plants.

Aspergillus can be manipulated to manufacture different types of enzymes by changing the type of aspergillus used or by changing the food source, the temperature, the humidity, the pH of, or the environment in which it is grown. Once mature, it is purified in water and is forced through a filtration system that leaves only the active enzyme. Then they are either maintained as a liquid, sprayed on a carrier (such as maltodextrine), freeze-dried or crystallized for use in foods, the production of alcohol and paper, cleaning detergents, nutritional supplements, and pharmaceutical drugs depending on the type of enzyme and the purpose of its use.

INHERENT BENEFITS
- The enzymes are vegetarian and vegan and can thus be consumed by everyone.
- They are the highest potency source, between 10 and 100 times more effective at digesting proteins, fats and carbohydrates per milligram than animal enzymes.
- Unlike animal sources, the health of the source in never questionable.
- The pH range is broad, making them active in stomach acid and throughout the rest of the body.
- Aspergillus can be manipulated to manufacture many different types of enzymes.

For all of the above reasons, plant-based enzymes are usually the first choice. In the following chapters we will see specifically how these enzymes are used digestively and therapeutically. Let's begin with the obvious...digestion.

Chapter Four
Phase One: Digestion

The most obvious use of enzymes to overcome a health issue or symptom is to use them to enhance digestion. Whether a person has indigestion, heartburn, acid reflux, gas, bloating, fatigue after eating, food cravings and the like, they can benefit from enzymes with their meals. Many of these symptoms are the result of inefficient digestion. Either animal or plant-based enzymes may be used; however the plant-based enzymes will provide the most support from a purely digestive approach. Although if the pancreas is inflamed, sluggish or diseased, this is a situation in which animal-based enzymes should be combined with plant-based enzymes as a regular part of a regimen. The animal source enzymes will fortify the pancreas, while the plant-based enzymes literally break down the food eaten.

Any of the symptoms noted above is an indication that an enzyme deficiency exists. For some reason the body cannot keep up with the demand placed upon it and the lack of digestive enzyme production will lead to common digestive complaints. If the symptoms are left unchecked, they can lead to more serious digestive problems and a host of diseases. Most people give little thought to the connection between their digestion and their overall health, but the two are inescapably intertwined.

Digestion and Health (Not Digestive Health)

By some estimates, 80 percent of the energy we use in our lifetime is used to digest the foods we eat! Think of it! Most of us eat three times a day, plus snacks. Completely digesting the average meal, from the time the food enters the body until the waste leaves, takes an average of three days. This means that every minute of every day we are digesting foods. Not a second passes that our digestive system is not working at breaking down foods, delivering nutrients and expelling waste.

What is the other 20 percent of our energy being used for? The answer is **everything** else. All the systems of the body including the immune, respiratory, reproductive, cardiovascular, nervous and muscular systems must share this remaining 20 percent. Isn't that amazing? The digestive system consumes four times more energy per day than all other systems combined.

Think of our body as a house in the Arizona desert in the heat of summer. Our digestive system is that big old air conditioner that runs all day to keep the house cool. When we get the electric bill, we have no doubt what appliance is costing us the most money; it's the one that uses the most energy. This is why digestive symptoms are often the first clue that something is wrong with our bodies.

Imagine, for example, that we have a problem with any of the other systems – immune, respiratory, reproductive, cardiovascular, nervous and/or muscular systems. The body is capable of handling problems, but it requires additional energy to do so. Where does that additional energy come from? It is "stolen" from the digestive system.

If we go back to the analogy of the house and the air conditioner, think of a digestive complaint or symptom as a circuit breaker that has tripped. The demand for electricity is too great for the circuit. Then we have to go to the breaker box and reset it, or in the old days replace a fuse. Sometimes it may require unplugging one appliance to run another. This is often the case when out of nowhere we start feeling heartburn (or any other digestive complaint). The body "turns off" one appliance (system) to run another. Of course it does not completely shut down; it just steals a bit of the energy normally used by the digestive system. For example, fewer digestive enzymes may be produced in order to provide additional white blood cells for immune support.

When your dog, cat or horse is sick, what is their normal response? They will *not* eat! You can put a pound of the world's best steak in front of a sick dog and you will get nothing more than a sniff out of him. The animal will instinctively preserve their energy to heal.

The more energy you can free up for the body to overcome an issue, the better your chances of overcoming that issue. When we're feeling sick, although we might not have an appetite, before long we realize it is already 2:00 in the afternoon and we haven't eaten anything yet. So we eat whatever is in the refrigerator, and think we're doing the right thing.

I am not suggesting we should always refrain from eating when we are sick, yet we need to consider eating foods that require little effort from the digestive system, and ones that are high in nutrition. Freshly juiced fruits and vegetables would be a great choice during times of illness. This is precisely why fasting is so beneficial to us and promotes healing. Imagine the energy you are giving back to the body by fasting! Calorie restriction provides the same opportunity; it allows more metabolic energy by reducing the demand for digestive energy. Many believe this abundance of metabolic energy is the mechanism for the success of the longevity diet. Raw food is also an option to return energy to the body. Since raw foods are full of enzymes, the enzyme content spares the pancreas from having to manufacture an excessive amount of digestive enzymes, providing more energy for other systems.

So what does this mean? It's quite simple actually. If you want to stay healthy or get healthy you need to free up as much digestive energy as you can spare. This will allow all of the other systems of the body to have energy they need to function properly. Fasting, restricting calories, eating more raw food or taking high potency plant-based digestive enzymes at every meal can facilitate this. Regardless of what a person is suffering from, the health professional that practices enzyme therapy will always start with digestion. That is probably the single most effective approach you can take to maintain or regain health.

Enzyme Potential

The term enzyme potential originated in Dr. Edward Howell's book *Enzyme Nutrition*. He describes that we are all born with a different potential for producing enzymes. Enzyme potential is the number of enzymes we can manufacture in our lifetime, either metabolic or digestive. The more digestive enzymes your body is forced to manufacture, the fewer metabolic enzymes your body will have the ability to produce. This in turn, leads to an enzyme shortage and the quickest way to disease and death. Dr. Howell illustrates this by suggesting we have a kind of enzyme bank account in which we are making constant withdrawals; the fewer withdrawals, the longer we live. Some have described this process by saying that the body will borrow metabolic enzymes to manufacture digestive enzymes and vice-versa.

Dr. Howell's theory has caused a lot of debate, however. When the theory suggests that there is a borrowing of one type of enzyme, such as metabolic, to serve digestively, the theory runs into problems. This is impossible because the specific nature of enzymes prohibits this exchange.

What we do know is that from puberty on, with each passing decade, our bodies make approximately 10-13% fewer enzymes than the previous decade. He references numerous studies in humans, insects and animals that all show a decrease in enzyme output with age.

If we consider that life requires energy and there is only so much energy to go around, the more energy you expend digestively, the less energy you will have systemically.

The Ideal Diet

The ideal diet is the one that allows you to live up to your life potential (the amount of time you remain alive), and provides the longest health span (the amount of time you remain healthy). We all have both a life and a health potential. Our life potential is how long we have the ability to live under ideal circumstances. Some of us have the potential to live to be 111, but because of the choices we make and the chances we take, we only live to be 70. Our health potential is how long we have the potential to stay healthy, vibrant and energetic. The ideal diet is one that allows you to maximize both potentials.

What does the ideal diet look like? One would need to eat nutritionally healthy foods in a raw state with few calories. It would also include regular fasting and cleansing. Those are not recommendations that many people would willingly choose, however. If you can't follow the ideal diet guidelines, then let's discuss the nearly ideal diet.

The Nearly Ideal Diet

The nearly ideal diet is a compromise of a combination of raw food, restricted calories, times of fasting and digestive enzymes. It's about decisions and choices. One decides what works, and sacrifices other choices. Nothing more, nothing less. The good news is that when the decisions we make are less than ideal, we can modify something we don't mind sacrificing to keep the "nearly ideal diet" in balance.

Let's look at this nearly ideal diet for a person who is in relatively good health. This person is not more than 15 pounds overweight, exercises fairly regularly, eats healthy foods most of the time, but wants to make some changes to increase their energy, lose those extra pounds and extend their "spans." (Life Span and Health Span)

All this person needs to do is:
- Reduce the size of his/her meals by 25%.
- Make at least one meal a day raw foods, or at least mostly raw.
- Take a high potency digestive enzyme with each meal.
- Fast with raw food once every three months for no less than five days.

Does that sound very difficult? The dividends it pays are truly amazing. The benefits to the average person include weight reduction (to your ideal, healthy weight), greater energy, reduced food cravings, more restful sleep, increased mental clarity, and most importantly an extension of both life and health spans. The beauty of these recommendations is that if you fail in one of the four suggestions, you can modify another.

If you have trouble eating a raw food meal every day, then you can increase the amount of digestive enzymes, reduce the meal size more than 25%, or fast every two months instead of every three months. The reason the nearly ideal diet is so adjustable is that every recommendation listed above contributes in a slightly different way to resting the digestive system and allowing the body to benefit from the additional energy.

Will this nearly ideal diet allow you to experience your full potential? No. We would have to be much more restrictive in order for that to occur. Yet by taking this nearly ideal approach, you will greatly increase your "spans." For the person in a health crisis, I recommend the nearly ideal diet, modified to maximize the amount of digestive energy preserved.

Phase one is all about digestive energy. Whether in a health crisis or just wanting to live a healthier life, this is where you must start! Ignoring these recommendations is like trying to hike up a mountain in flip-flops. You have to start with the basics. In Chapter Eight I discuss the importance of maintaining a proper pH and bacterial balance as part of a healthy regimen. This too, should be considered as part of the basics.

Chapter Five
Phase Two: The Use of Therapeutic (Systemic) Enzymes

The difference between digestive enzymes and therapeutic enzymes (or what some people call systemic enzymes) is timing. Taken at different times, in different situations, each of these enzymes can have different effects. When taken at the beginning of a meal, the enzymes will assist the digestive process. When taken apart from meals on an empty stomach, they are considered to be therapeutic. As mentioned in the previous chapter, one must start with digestion in enzyme therapy. Once this is adequately addressed, the enzyme therapist tries to determine which category of enzymes will assist their client in overcoming the most obvious deficiency to produce the fastest results.

The use of therapeutic (systemic) enzymes can be started along with digestive enzymes, provided digestion has been addressed. If an individual is not taking digestive enzymes, eating raw food, restricting calories or intentionally fasting, then Phase Two becomes much less effective, especially if undergoing a health crisis. For therapeutic enzymes to have a healing effect, it is imperative that there is not a large demand for digestive energy 24 hours a day, 7 days a week.

The use of enzymes therapeutically is similar to using them digestively. Though there are companies and people who would like you to believe that there is a difference, there is not. The only difference is the **timing of dosing**. Both approaches can cause a systemic response. When used therapeutically on an empty stomach, the enzymes are not digesting food, they are absorbed into the blood and can have a therapeutic effect on different systems of the body. On the other hand, even digestive enzymes taken with food can be therapeutic (systemic) if more than necessary to digest a meal are consumed. The extra enzymes are then absorbed and utilized throughout the body.

Why would a person want to consume enzymes on an empty stomach so that they are absorbed into the blood? Think back to the different enzymes we discussed in Chapter One. Metabolic enzymes make biochemical reactions possible in all living cells; a deficiency of these precious proteins is what results in disease. The therapeutic or systemic use of enzymes is designed specifically to support this deficiency. Each enzyme category can address specific deficiencies; by correcting the deficiency, the health issue is then corrected.

All disease is the result of a deficiency, a lack of something essential to health, which results in sickness and pain. When the body is no longer deficient, it is no longer sick! Another way to explain this principle is that all disease is the result of an imbalance; by bringing the body back into balance one can improve one's health.

Why We Get Sick

Traveling on airplanes is a great place to get sick. When we sit on a plane, we all breathe the same recycled air. If someone happens to have a cold or virus that is airborne, we are exposed to it. Please notice that I didn't say, "We are going to get sick." I said we will be exposed to the virus. Though everyone on the plane is exposed, not everyone will get sick. Isn't that interesting? Why would some people who are exposed to an airborne virus get sick and others will not? The simplest answer is we are all different. But what makes us different? Some of us have something that the others don't; those who are deficient will get sick, while others with no such deficiency remain healthy.

To really drive this point home think about the deadly virus HIV. The fact is that thousands, if not millions have been exposed to HIV and have not contracted the disease. Why? The short answer is that some people lack the metabolic enzymes needed to maintain their optimum health; when they are exposed to a virus or bacteria, they'll get sick. Though this may seem oversimplified, think of how remarkable this is. Metabolic enzymes play such a vital role in health, yet few have ever heard of them, much less understand the essential role they play in maintaining health.

The reason why so few know or write about metabolic enzymes is because the metabolic enzymes we lack cannot be placed in a pill. Since many feel there is little value in talking about something that cannot bottled and sold, it simply is not discussed. What these people fail to grasp is that even though we may not be able to replace the enzymes we are specifically deficient in with enzyme supplements per say, we can supply the body with what it needs to support the deficiency and thereby assist it in getting back to manufacturing the lacking enzymes.

Therapeutic Enzyme Supplements

There are very few companies that specialize in enzyme

supplements. A few that do are Enzymedica, Theramedix, Tyler, Enzyme Research, Transformation and Mucos Pharma. One would think that a company with the name Enzymatic Therapy would specialize in enzymes, yet they don't. Although they make excellent products, only a small percentage are enzyme based. In fact, of the companies mentioned above, only a few provide supplemental enzymes exclusively. Most have an extensive line of vitamins, minerals and herbal blends combined with enzymes. One of the reasons why Enzymedica and Theramedix are among the exceptions is because the market is relatively small for such products. Since so few individuals understand the concept of enzyme therapy, only a small number of consumers will ever purchase a product made up exclusively of enzymes.

It is important to note that none of these companies provide metabolic enzyme products. Instead, they use plant-based, plant and animal enzymes in their formulas designed to be taken apart from food to be absorbed in the bloodstream. (Remember that metabolic enzymes are manufactured by the body. They are produced in all living cells and are responsible for nearly every biochemical reaction that takes place in every cell.) Though plant-based, plant and animal enzymes may not function as metabolic enzymes, they have been shown to have an effect on metabolic function.

For example, white blood cells, which play a key role in immune function, are full of enzymes. They are approximately twice the size of red blood cells and circulate through the body consuming pathogens such as viruses, bacteria, fungal forms and parasites, removing them as a threat. In order to perform this task, our white blood cells must contain thousands of enzymes. The primary category of enzymes that are contained within a white blood cell are proteases, which break down protein. Since everything a white blood cell must consume is either protein (bacteria, parasites) or protected by protein (viruses, cancer cells) protease is absolutely necessary for proper immune function. Imagine what would occur if we became deficient in the proteases that make up the white blood cells in our immune system! It is not hard to fathom that the result is smaller, less active white blood cells and an under-active immune system. To support this deficiency, proteases can be supplied orally to assist the body in addressing this need. Of course if the proteases are consumed with food, they would be utilized digestively and therefore would have little impact on the specific need of the immune system. So it is recommended that the enzymes are taken on an empty stomach, which generally means a half hour before or two hours after food.

The question then becomes, how do enzymes from a plant-based, plant or animal sources provide assistance when a metabolic protease deficiency exists. Keep in mind that enzymes are very specific when it comes to what they can do. There are two possible answers. The first is that the proteases taken as a dietary supplement act as raw material by bone marrow and the lymph system to manufacture additional white blood cells.

The second is that the protease begins digesting dead and damaged cells throughout the body, reducing the need elsewhere for protease. This gives the body a rest, and the proteases that would have been produced for this purpose can now be employed for immune function processes. In either case, the result is the same: the consumption of proteases on an empty stomach produces more active white blood cells to support healthy immune function.

This is just one example of how the second part of enzyme therapy works. Once digestion is addressed, the enzyme therapist tries to establish what category of enzymes will pursue the most obvious deficiency to produce the fastest results. After this is determined, the enzymes are first taken on an empty stomach to elicit a metabolic response. Now let's look deeper into the specific categories of enzymes and how they can be used therapeutically.

Chapter Six
Specific Enzymes in Therapy

As mentioned in the previous chapter, the enzyme therapist tries to determine what category of enzymes will address the most obvious deficiency, producing the fastest results. The reason I said **most obvious** is because there are often multiple issues. For example, let's say Sue is suffering from heartburn, arthritis and fatigue. Her diet is 90% cooked and processed foods; she exercises minimally. We always start with digestion first, so the initial recommendation for Sue is to take digestive enzymes with every meal while increasing the amount of raw food she consumes daily.

I would personally recommend fresh fruit in the morning, with a raw salad for lunch or dinner. I would also suggest she try to cut her overall caloric intake by 25% by making portions smaller and snacking on carrots, celery, apples, pears and other fresh fruits and vegetables between meals. This alone will contribute to an enormous release of energy that has previously been designated to digesting a diet of processed foods. Sue may be able to overcome the fatigue and resolve her digestive complaints with just these simple suggestions.

However, arthritis with its painful symptoms is still a

concern for her. Here is where we need to get specific with enzymes between meals. In order to make a recommendation at this point, what needs to be understood is the fundamental effects that enzymes have on the body. She will need an enzyme blend that is primarily proteolytic (one that breaks down protein) since proteases are known to reduce inflammation on many levels. The proteolytic formula would contain a high percentage of bromelain, papain and protease. Other enzymes that may additionally prove useful are nattokinase, catalase and serratiopeptidase (See Appendix). In fact, all of these enzymes have been shown to increase circulation and reduce inflammation. Though this may sound a bit confusing, let's attempt to simplify it by outlining the most popular enzymes and their primary uses in alphabetical order. (Included in the outline below are the units by which the enzyme's activity is measured; there are several different units of measurement and this will be covered in more detail in Chapter Seven: Determining Enzyme Potency.)

Enzymes and Their Uses:

Alpha-Galactosidase
- Breaks down carbohydrates, such as raffinose and stachyose
- Especially helpful with raw vegetables and beans
- Measured in GAL (Galactosidase Units)

Amylase (Carbohydrase)
- Breaks down carbohydrates, such as starch and glycogen
- Regulates histamine when taken on an empty stomach

- Reduces food cravings
- Increases blood sugar
- Available from different sources and can be blended to increase potency
- Measured in SKB and DU (Dextrinizing Units)

Bromelain
- Breaks down protein
- Most beneficial as an anti-inflammatory
- Measured in GDU (Gelatin Digesting Units) and FCCPU

Catalase
- Acts as an antioxidant by breaking down hydrogen peroxide into water and oxygen
- One of the most potent antioxidants, found in nearly every cell of the body
- Measured in Baker Units ("Baker" invented the Assay)

Cellulase
- Breaks down cellulose and chitin, a cellulose-like fiber found in the cell wall of *Candida*
- Helps free nutrients in both fruits and vegetables because of its action on the cell wall
- Available from different sources and can be blended to increase potency
- Measured in CU (Cellulase Units)

Diastase [See Maltase]

Glucoamylase
- Breaks down carbohydrates, specifically polysac-

charides, or long chains of carbohydrates
- Measured in AGU (Amyloglucosidase Units)

Glucoreductase
- Breaks down blood glucose
- Measured in GRU (Glucoreductase units)

Hemicellulase
- Breaks down carbohydrates
- Especially helpful with polysaccharides found in plants
- Measured in HCU (Hemicellulase Units)

Invertase (Sucrase)
- Breaks down carbohydrates, especially sucrose and maltose
- Measured in IAU (Invertase Active Units)

Lactase
- Breaks down lactose (milk sugar)
- Used to treat lactose intolerance
- Measured in LacU (Lactase Units)

Lipase
- Breaks down lipids and improves fat utilization
- Helps reduce cholesterol
- Supports weight loss
- Supports hormone production
- Supports gallbladder function
- Available from different sources and can be blended to increase potency
- Measured in FCCFIP

Maltase (Diastase, Malt Diastase)
- Breaks down carbohydrates, malt and grain sugars
- Breaks down complex and simple sugars
- Measured in DP (Degrees of Diastatic Power)

Mucolase
- The non-crystalline form of seaprose
- Breaks down mucus
- Helpful for congestion and sinus infections
- Measured in milligrams and MSU (mucolase units)

Papain
- Breaks down protein
- Most beneficial as an anti-inflammatory
- Measured in FCCPU (Papain Units)

Pectinase
- Breaks down carbohydrates, such as pectin found in many fruits and vegetables.
- Measured in AJDU (Apple Juice Depectinizing Units)

Phytase
- Breaks down carbohydrates
- Especially helpful in breaking down phytic acid found in the leaves of plants
- Helps with mineral absorption
- Measured in PU (Phytase Units)

Protease
- Breaks down protein
- Bonds with alpha 2-macroglobulin to support immune function when taken on an empty stomach

- Reduces inflammation and increases circulation
- Available from different sources and can be blended to increase potency
- Measured in HUT (Hemoglobin Units in a Tyrosine Base)

Seaprose
- Breaks down mucus
- Helpful with congestion and sinus infections
- A crystalline (more concentrated) form of mucolase
- Measured in mgs (milligrams)

Serratiopeptidase
- Anti-inflammatory
- Measured in units (This is a relatively new enzyme and does not yet have an abreviation; it will simply say 10,000 units on the bottle.)

Sucrase [See Invertase]

Xylanase
- A type of Hemmicellulase
- Breaks down soluble fiber rather than insoluble fiber
- Measured in XU (Xylanase Units)

Categories of Enzymes
The most widely researched category of enzymes are those that break down proteins. They are often called proteolytic enzymes or just proteases. When proteolytic enzymes are consumed with food, they assist in breaking down proteins. When consumed between meals, they are absorbed into the blood to assist with immune imbalances,

heavy metal toxicity, inflammatory conditions, circulatory disorders, skin problems, constipation, water retention, inappropriate blood clots, heart disease, stroke, cancer and the like. The use of proteases for these conditions is the second most popular use of enzyme therapy, after digestive applications. The logical reasons for their popularity are their many applications.

The one thing that needs to be understood about illness, aside from the fact that it is the result of an enzyme deficiency, is that the cause of the illness is in some way related to protein. For example, cancer cells are surrounded by a protein coat; blood clots that cause stroke and a high percentage of heart attacks are made up of a protein called fibrin; pathogenic bacteria and parasites are comprised of proteins; fungal forms such as *Candida* are made up of a protein nucleus surrounded by a chitin shell. Viruses are enveloped by protein. An amazing coincidence isn't it? In order for us to stay healthy we need to be well equipped to overcome protein invaders in the body that will make us ill if left unchecked.

Here's a surprise: we are well equipped. To quote Dr. Ellen Cutler, M.D: "Our immune system is overbuilt for success."[1] Unfortunately, we can make it under-equipped by overeating cooked and processed foods. Since digestion always takes the highest precedence, our body will sacrifice energy utilized for immune function to digest the foods we have eaten. If those foods are devoid of enzymes, a greater need arises. After years of repeating this pattern, the body can no longer keep up; a shortage then occurs, which eventually takes its toll. We soon find ourselves deficient in the enzymes assigned to stand guard and take out any invader that poses a threat. These enzymes are not just floating around looking for an invader—no, enzymes don't look for anything. But they are what make white blood cells effective in overcoming dangers. A protease deficiency allows bacteria, viruses, and all other pathogenic processes to have there way with us. Therefore, the last thing we want is a protease deficiency.

Protease, The Immune System And Alpha 2 Macroglobulin

When we introduce oral proteases on an empty stomach, they have the ability to act upon the proteins that can make us sick. Taken by mouth on an empty stomach, proteases are readily absorbed into the mucosa cells of the intestine and into the blood. Once there, they join a biochemical secreted by white blood cells known as alpha II-macroglobulin (A2M for short). A2M can be thought of as an escort with two purposes; the first is to protect the protease from removal from the bloodstream, and the second is to take it where it is needed. At this point, it would be good to take a quick look at the immune system to further our understanding of this amazing process and the connection it has to proteolytic enzymes. [This is discussed in greater detail in Chapter Ten.]

There are several types of cells in the immune system that play one of two roles. They either identify what is harmful or they remove it. One type of cells are T-cells that circulate through the body identifying harmful bacteria and viruses. Once identified, T-cells chemically mark pathogens so other cells of the immune system know they are "wanted." (Think of this as placing a flag on them.) Next, several types of white blood cells (and there are several kinds) are sent out looking for flags. Upon finding a flag-marked pathogen, it engulfs and digests it.

One type of white blood cell is called a macrophage,

which literally means "big eater" in Greek. The white blood cell then communicates with other cells of the immune system by releasing biochemicals that send a message that an invader has been found and taken care of. One of the many biochemicals released during this process is A2M.

While science does not completely understand the exact role of each immune system chemical (many serve more than one purpose), the ability of A2M to bond with protease and escort it throughout the body is well known. Remarkably, A2M appears to have the same ability that white blood cells have for identifying what does not belong. Once an intruder is identified, the A2M exposes the protease to the protein-based invader and then digestion of that pathogen begins. That process makes it possible to supplement the immune system with orally taken protease. We can actually supply the body with a supplement that becomes a part of the immune system!

There are several different types of protease, derived from four main forms:

Fungal (protease, catalase, and seaprose)

Bacterial (serratiopetidase or serrapeptase, nattokinase)

Plant (bromelain, papain and ficin)

Animal (pancreatin, trypsin and chymotrypsin)

Lipases

The second most researched group of enzymes in the world are lipases. These enzymes are lipolytic, in that they break down or disengage fat. Lipases are found in both fungal and animal enzymes. They are one of the simplest enzymes to understand and one of the easiest to recommend, as they are effective for issues related to fats. For example, obesity can be thought of as a deficiency in lipase. One study showed that 100% of clinically obese (over 30% their ideal body weight) individuals are lipase deficient.[2]

Fat is found in our cells, skin, blood (as HDL and LDL cholesterol), and the sheath that surrounds our nerves. Fat also plays a crucial role in hormone production. Additionally, the fat-soluble vitamins A, D, E and K require lipase to benefit the cells. For all of the above reasons, lipase is recommended therapeutically for high cholesterol, obesity, high triglycerides, heart disease, hormonal imbalances, nerve problems, fat-soluble vitamin deficiencies and skin problems such as eczema and psoriasis.

Amylases

The third category of enzymes are amylases that digestively break down carbohydrates. This category is a bit more complicated because technically speaking amylases break down sugars, fiber and complex carbohydrates, all of which are carbohydrates.

Basic amylases help to break down complex carbohydrates such as fruits, vegetables and legumes. Therapeutically they have been shown to regulate histamine, which is produced by cells in the body when a perceived invader is recognized. Histamine is responsible for the common allergy symptoms many people experience when the pollen count in the air is high. It is difficult to say exactly how amylase is involved in this process, but it most certainly is.[3] If you have any doubts at all simply try taking amylase on an empty stomach next time your hay fever flairs up.

Amylase is also good at RAISING blood sugar. If you have sugar cravings, food cravings and low blood sugar, amylase may help tremendously. The carbohydrate-digesting amylases from fungal (plant-based) sources are: amylase, maltase, glucoamylase, alpha-galactosidase, hemicellulase, xylanase, pectinase and phytase.

The amylase enzymes that digest sugars are involved in breaking down sucrose, lactose and maltose. When these sugars are not properly digested, people often exhibit symptoms that include depression, moodiness, panic attacks, manic and schizophrenic behavior, severe mood swings, abdominal cramps, diarrhea and environmental sensitivities. The sugar-digesting amylases from fungal (plant-based) sources are: sucrase, lactase and maltase.

The amylases that digest fiber are called cellulases, which is the only enzyme our body does not manufacture, although the friendly bacteria in our large intestine do produce cellulase for our benefit. If the microflora (bacteria) needed in our large intestine is out of balance however, the resultant lack of cellulase can contribute to many symptoms associated primarily with a condition called Candidiasis. Candidiasis is the collection of symptoms caused by *Candida albicans*, a yeast-like fungus that normally lives in healthy balance in the body. It is found mostly in the intestines, genital tract, mouth, and throat. When this balance is upset, infection results when the fungus travels to all parts of the body through the bloodstream. In the mouth, it is called thrush; in the vagina, it is called vaginitis (yeast infection). The symptoms which effect both men and women include new allergies to foods, fatigue, poor digestion, gas, heartburn, sugar cravings, irritability, frequent headaches, poor memory, dizziness, recurring depression, vaginal infections, menstrual difficulties, prostatitis, urinary tract infections, hay fever, postnasal drip, habitual coughing, sore throat, athlete's foot, skin rash, psoriasis, cold extremities, and arthritis-like symptoms.[4]

In individuals with *Candida albicans* overgrowth there is often a large amount of undigested fiber in the large intestine. It is believed that the mucus the body naturally produces due to the difficulty of digesting this fiber may protect the *Candida* from our body's natural overgrowth prevention mechanisms. Since cellulase breaks down fiber, it can resolve this problem by removing both the fiber and the mucous so our body is able to achieve a state of balance.

Another aspect of why cellulase can help therapeutically with *Candida* is how it can specifically remove the overgrowth. The *Candida* cell contains some fungal cellulose or chitin.[5,6] Chitin is a polysaccharide or carbohydrate, which is structurally very similar to cellulose. For the same reasons cellulase helps remove undigested fiber in the colon, it has been used in breaking down this "cellulose-like" chitin. It is important to subsidize cellulase with a probiotic (a friendly bacteria such as *L. acidophilus*) when trying to correct this problem. This is addressed further in Chapter Nine.

References:
1 Cutler, Dr. Ellen. *Micro Miracles: Discover the Healing Power of Enzymes.* Holtzbrink Publishers/ 2005.

2 Roitt I, Brostoff J, Male D. Cells involved in the immune response. In: Roitt I, Brostoff J, Male D (eds). *Immunology*. St. Louis, Mo: The CV Mosby Co; 1985: 2.1-2.16.

3 Desser, L., Rehberger, A., "Induction of tumor necrosis factor in human peripheral-blood mononuclear cells by proteolytic enzymes," *Oncology* 47:475-77 (1990).

4 *Nutrition Science News*, May 1997.

5 The Penguin Dictionary of Biology.

6 Oxford Dictionary of Biology.

Chapter Seven
Determining Enzyme Potency

The potency of enzymes are not measured the same way other nutritional supplements are measured. Potency is defined as, "the quality or state of being potent; ability to effect a purpose; power; strength; energy; capability; efficacy; influence." The determining factor of an enzyme product's potency is the "effect" it has on proteins, fats and carbohydrates. In other words, the quantity of food that an enzyme can break down or digest determines its potency.

This method of measurement may differ from what most people are accustomed to. For example, when comparing two vitamin C products, the average consumer will typically compare the price and number of milligrams per tablet of one vitamin C product with another. When it comes to enzymes however, it is not quite as simple. One must consider three factors when determining enzyme potency.

Below is a copy of the supplement facts box from the best selling digestive enzyme product sold in health food stores. It is called "Digest Gold" and is manufactured by a company called Enzymedica. Observe how the ingredients are measured.

Nutritional Information	
Serving Size: 1 capsule	
Capsules Per Container: 90	
Amount Per Capsule:	
Amylase Thera-blend	23,000 DU
Protease Thera-blend	80,000 HUT
Maltase	200 DP
Glucoamylase	50 AG
Alpha-Galactosidase	450 GALU
Lipase Thera-blend	3,500 FCCFIP
Cellulase Thera-blend	3,000 CU
Lactase	900 ALU
Beta Glucanase	25 BGU
Xylanase	550 XU
Pectinase / Phytase	45 endo-PGU
Hemicellulase	30 HCU
Invertase	79 INVU
L. Acidophilus	250 million CFU
This product contain(s) no fillers.	

In this example you can see that the name of the enzyme is listed in the left panel and measurement of activity is listed in the right panel followed by two to four letters. The word *Thera-blend*, which follows the name of four of the enzymes, is Enzymedica's way of describing that more than one enzyme is used. This will be covered in greater detail toward the end of the chapter.

What initially confuses most customers are the letters shown after the measurement of activity. On the Nutritional Information panel, amylase has 23,000 DU, while protease has 80,000 HUT. These letters are abbreviations for assays used to measure the "active units" of the enzyme. DU stands for Dextrinizing Units, while HUT is an abbreviation for Hemoglobin Units in a Tyrosine base. The reason this is necessary is because enzymes are not measured by weight and thus a measurement in milligrams (mg) or international units (IU) would not describe the true potency of the product.

1) The active units are the most commonly used measurement to determine potency since it identifies how active the enzyme is. An "active unit" is a measurement that describes how much of a given food an enzyme has the potential to break down. For example, 1,000 active units of one type of lipase (the enzyme that digest fats) has the potential to release 1,000 bonds of essential fatty acids from olive oil per minute. When the lab technician tests the enzyme for its activity though, they do so in a very controlled environment in which the temperature and pH are closely regulated. To measure the activity of the lipase mentioned above, a lab technician will expose a measured amount of the enzyme to a fat (olive oil), in a pH of 7 at a temperature of 86 degrees to determine its "active units." The measurement would tell you that the lipase has the ability to liberate 1,000 bonds of fatty acids per minute which will be labeled 1,000 FCCFIP on the label. However, since the enzyme is measured using one type of fat in a very specific pH, it is only an approximation of the potential to digest other types of fats in other pH's or temperature.

2) In the test mentioned above one of the key factors in determining the potency of an enzyme is the pH in which the enzyme was tested. pH is simply a measurement of acidity and alkalinity. The higher the number, the more alkaline

the substance; the lower the number, the more acidic the substance. pH can range from 0 to 14 while 7 is neutral (the middle of the scale). The pH range of enzymes is an important measure of potency since it defines how long it will work in the body. All enzymes have an optimal pH and pH range. When it comes to determining the potency of a product, it helps to know that one of the manufacturer's considerations was the pH in which the enzyme works best. This is a very important consideration since different portions of the body function in varying pH levels. The pH of the stomach averages between 2 and 3 (very acidic), while the small intestine is alkaline at a pH of 8; the blood remains slightly alkaline at just over 7 on the pH scale. So, if an enzyme product contains a protease that has an optimal activity of 3, it will work well at digesting protein in the acid environment of the stomach, yet may not work at all in the alkaline environment of the small intestine or the blood (See Chapter Eight for more information about pH).

The following image illustrates how the optimal pH ranges of enzymes will often vary.

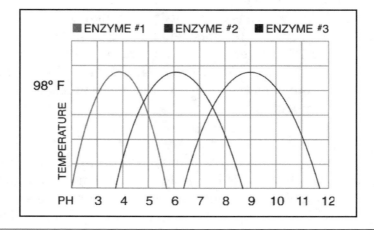

3) The third consideration is the number of enzymes used per category. Protease, lipase, amylase and cellulase are actually categories of enzymes. Not all proteases can digest all proteins, nor can all lipases digest all fats. So the more proteases in a protease blend (such as the Enzymedica "Thera-blend") the more protein it will be able to break down. The combining of enzymes within categories (such as multiple proteases, lipases, amylases and cellulases) will digest or break down more protein, fat, carbohydrates and fiber, respectively. This blending also increases the range of pH the enzymes work in and illustrates how one enzyme will break less bonds and is less potent than a multiple enzyme blend.

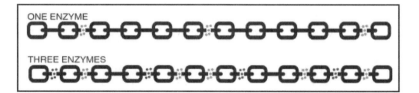

From this discussion, it is apparent that there is a lot to the science of enzyme potency. Because of this, it may be difficult for the consumer to be certain of value and efficacy when purchasing enzyme products. Therefore, I recommend purchasing enzymes from a company that specializes in the manufacturing of such products. Few companies have the expertise needed to produce properly formulated enzyme products that truly perform as intended.

How much actual protein, fat and carbohydrates can our enzymes digest?

This is a complicated question but note that every protein, carbohydrate and fat is different. For example:

20,000 HUT of protease can break down 225 grams of dairy protein in one hour. Beef or soy would be different.

30,000 DU of amylase can break down 100 grams of potato starch in one hour. The starch found in a carrot would be different.

1,000 FCCFIP of lipase can break down 5 grams of vegetable oil in one minute. The fat found in beef or avocado would be different as well.

A Word About Value

When determining value you will no doubt try to compare enzyme supplement labels. Chances are the only thing you will find of help are the active units listed. Because the blending of enzymes is so critical to potency, you cannot assume that the higher the active units the better the value. If you are comparing a product that costs the same price and has the same number of capsules per bottle, note whether or not one contains multiple enzymes per category.

As an illustration, one product that contains multiple enzyme blends may contain 80,000 HUT's of protease whereas a competing product may list 100,000 HUT's. Despite the difference in active units, the 80,000 HUT product will be more potent since the blended product will work longer in the body and ultimately be able to break more bonds than the other product. Which is a better value? In this example the one with less active units but more enzymes would be a better value!

Chapter Eight
The Importance of Proper pH

The term "pH" defines a scale that assigns a number to indicate acidity and alkalinity. It stands for "potential of hydrogen" and defines the relative amount of hydrogen ions in a particular solution. The more ions there are, the more acid the solution. The fewer the ions, the more alkaline (basic) the solution is. pH is measured on a scale of 0 to 14 with seven being neutral. The lower the pH number, the more acid it is. The higher the number, the more alkaline. For example, a pH of 3 is more acidic than a pH of 5. A pH of 9 is more alkaline than a pH of 6.

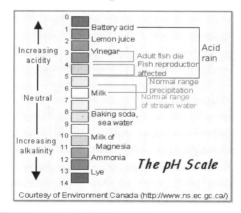

Courtesy of Environment Canada (http://www.ns.ec.gc.ca/)

Homeostasis

One definition of homeostasis is the ability or tendency of an organism or a cell to maintain internal equilibrium by adjusting its physiological processes. This is a truly remarkable process in which the body adjusts according to the need. An example might be the production of the hormone epinephrine (adrenaline) when the body detects danger. This response increases heart rate, breathing and energy to better handle the perceived danger and decrease the risk of harm.

pH balance within the body is part of this homeostasis process. The body must maintain a specific acid-alkaline balance to survive. However, this is much more critical in certain parts of the body than others. For instance, pH within the blood cannot vary much at all or death will quickly follow. On the other hand, the digestive system should maintain a specific acid-alkaline balance, but it can vary widely. It is out of balance more often than not in most adults. This is a good example of the body's ability to prioritize. Although it will do all it can to maintain the specific pH required in the blood, the body is often unable to meet the same specifications throughout the digestive tract. If it did, some other part of the body (perhaps a more critical organ or system) may be adversely affected due to neglect.

What is Normal?

As humans, a normal pH of all tissues and fluids of the body (except the stomach) is slightly alkaline. The most critical pH is in the blood. All other organs and fluids will fluctuate in their range in order to keep the blood at a strict pH between 7.35 and 7.45 (slightly alkaline). In addition to balancing other systems, the body makes constant adjustments in tissue and fluid pH to maintain this slightly alkaline pH range in the blood.

Too much acidity or alkalinity in the body can have far-reaching consequences. **For example, if the blood becomes too acidic:**

1) It takes some of the alkaline-forming elements from the enzymes in the small intestine to stay balanced. The small intestine then becomes too acidic to digest foods optimally. The pancreas, gallbladder and liver are then forced to make up for this deficiency in order to metabolize foods properly. This has a direct bearing on metabolic (or systemic) enzyme production, which results in lowered immune function, fatigue, hormonal imbalances, absorption and digestive problems, and many other problems as well.

2) Calcium starts leaching from the bones, since it is the most alkaline of the minerals. This occurs as the body struggles to maintain the homeostasis of the blood pH. This condition can lead to reduced absorption of supplemental minerals and bone density problems including osteoporosis.

3) Insulin levels increase and fat is stored instead of being metabolized. This is due to the body's mimicking what happens when malnutrition or starvation sets in. Interestingly, the body increases in acidity when malnourished; as a safety mechanism, insulin is over-produced, so that all available calories are stored as fat for future use. As a result, weight gain is common and weight loss becomes more difficult.

4) Electrolyte imbalances occur which have a direct bearing on the "fluid transport system." Electrolytes are important since they maintain the electrical voltage existing around all cells, especially critical around the cells of the heart, nervous system, and muscles.

Similar problems may occur if the body becomes too alkaline, though this is much less likely. Obviously maintaining the proper pH in the blood, digestive tract, tissues and fluids is essential in order to support optimal health.

What are Some Symptoms of an Imbalanced pH?

Symptoms include, but are not limited to, acid reflux, indigestion, weight gain, difficulty losing weight, poor metabolism, mineral deficiencies, constipation, fatigue, brain fog, frequent urination, hypoglycemia, hormonal imbalances, sore muscles and so on.

Some of the factors that contribute to this imbalance include stress, environmental pollution, too little or too much exercise, and the most important factor, diet. The more acid-forming foods we eat, the more acidic we become. The more alkaline-forming foods we eat, the more alkaline we become. **Generally speaking, fruits and vegetables are more alkaline-forming, while meats, sugar, caffeine, beans, dairy and grains are more acid forming.**

Notice the term "generally speaking" in the previous sentence. Gabriel Cousens, in *Conscious Eating* mentions the complexity of this topic. His research has shown that about 30% of the people he counseled responded exactly opposite to that. In other words, the fruits and vegetables they ate made them more acidic. So we must be willing to take the time to determine individually for ourselves, what foods will work best for *us*.

One of the best ways to determine whether or not a proper pH is being maintained is to test the urine. This is an accurate way to almost directly assess the pH of the blood. The kidneys are the main organ system that balances the pH of the blood. If we become too acidic, our kidneys will eliminate and remove acid through the urine. This is part of the homeostasis mechanism mentioned earlier and contributes to keeping blood alkalinity in balance. The urine will always reflect what the body is getting rid of in order to maintain homeostasis. Urine then, is an excellent indicator to determine whether our diet is too rich in foods that make us either acidic or alkaline. The more acidic the urine is, the more acid the body is trying to expel in order to maintain its alkalinity.

Although many people believe that the best time to test urine pH is the first thing in the morning, it is highly recommended to test this pH over a period for 24 hours. It is much more accurate to track pH throughout the day and then calculate the average.

In order to optimally benefit from this exercise, it is suggested to also keep a log of the foods consumed. This will help quickly determine which foods and drinks contribute to becoming more acidic or alkaline. Then you will be able to make adjustments to stay in balance through dietary changes.

The optimal urine pH is between 6 and 7. If your average is below 6, you are too acidic. If your average is above 7, you are too alkaline. In either case you should immediately consider lifestyle changes to support the proper pH for your body.

A Word About Saliva

It is often recommended that saliva is a good indicator of pH balance. Nevertheless, this method has been found to be very unreliable. Most people test alkaline when testing their saliva alone, while most people test acidic when testing only their urine. The fact is that when a person's saliva tests acidic, it indicates an extreme imbalance. Most experts agree that the pH of saliva is an indicator of alkaline reserve. This is because the saliva naturally contains minerals. When those minerals are not present, there is an alkaline reserve or mineral deficiency. The body does not create alkalinity; rather it has an alkaline mineral reserve based on the dietary intake of alkaline foods.

The healthy pH of saliva tested first thing in the morning or on an empty stomach is between 6.2 and 7.2. After a meal it should become even more alkaline. One theory is that if the pH is between 5.8 and 6.2, the body is too acidic with little alkaline reserve left. If the morning pH is below 5.8 with no rise after meals, there is no alkaline reserve left and the body is extremely acidic.

Helpful Lifestyle Changes

Diet is probably the most important change in creating a healthy lifestyle. Avoiding the over-consumption of meat, alcohol, soft drinks, caffeine, coffee, most nuts, eggs, vinegar, sauerkraut, ascorbic acid (vitamin C), pasteurized milk, cheese, white sugar and medical drugs is optimal. Adding additional servings of ripe fruit, vegetables, soybeans, bean sprouts, water, raw milk, onions, figs, carrots, beets, miso, and mineral supplements to your diet is extremely beneficial.

It also helps to reduce anxiety when possible and include moderate exercise in your daily regimen. Strenuous exercise can actually contribute to an acidic environment in the body because of the increased production of lactic acid. Though I do not suggest strenuous exercise is bad, it should be a consideration if one is having difficulty achieving an acid-alkaline balance.

Supportive Supplements

There are a number of products sold in health food stores that may prove supportive in balancing pH, though caution should be exercised. In my opinion most products formulated to help raise alkalinity are often too alkaline.

I have tested some of the most popular products sold for this purpose and the majority of products average 10 on the pH scale. Keep in mind that the most alkaline part of your digestive system is the small intestine, which averages 8 on the pH scale. The problem with putting something in your gut that is so alkaline is that it neutralizes stomach acid. The stomach averages between 2 and 3 on the pH scale and when this pH changes, the body will adjust to maintain balance. When neutralized, the body will work overtime trying to maintain the naturally acid environment of the stomach. This mechanism creates a demand to constantly produce more acid, which is obviously self-defeating when trying to become more alkaline.

This common response is why the person who consumes "antacids" for the symptom of heartburn will need more and more as time goes on. Whereas at first, a couple tablets did the trick, now at least five or six are required to experience any relief at all. This vicious cycle continues until the person begins taking prescription "H2 Blockers" or "proton pump inhibitors" that literally "shut off the acid faucet" so to speak. This of course, has other detrimental consequences that include poor protein digestion, ulcers and gas to name just a few.

The solution to the dilemma is actually quite simple. In order to raise alkalinity, you can take a product that does not exceed the natural pH of the small intestine and does not release in the stomach. This can be achieved by using a formula that maintains a pH of 8 with "enteric coating." The enteric coating protects the capsule (or tablet) from breaking down in the naturally acidic environment of the stomach, and instead, releasing in the small intestine where the ingredients can restore balance to this vital section of the intestinal tract and ultimately to the whole body. The beauty of this is that if the person is too alkaline (meaning the pH of the small intestine is above 8) the same formula can contribute to reducing alkalinity. This way it truly balances pH, rather than just raising alkalinity.

The only companies that I am aware of that go to such great lengths to create moderately alkaline, enteric-coated products are Enzymedica and Theramedix. The Enzymedica product is called "pH Basic" and the Theramedix product is called "pHB." Below are listed some of the active ingredients that are helpful in such formulas.

1) A **mineral blend** that consists of the same minerals often depleted in an acid environment and those that make up the electrolyte mineral ratio. These include potassium bicarbonate, sodium bicarbonate, and magnesium citrate.

2) An **organic super food** (greens) such as hydrilla, which is naturally alkaline. It is nature's most potent source of calcium, and one of the richest plant sources of many trace minerals and amino acids.

3) An **enzyme blend** formulated to promote the uptake of the minerals, herbs and greens.

4) The **herbs** marshmallow and papaya leaf extract help soothe the common symptoms of acid reflux that are often associated with high acidity.

5) A **delivery system** that bypasses the acid stomach and goes straight to the small intestine. This is important because if you neutralize the stomach acid, the body will manufacture more acid to maintain a pH balance.

Proper Digestion is Essential

In addition to dietary intake, in order to maintain optimum pH balance, it is important to ensure proper digestion and assimilation of the foods consumed. One of the best ways to stay balanced is to take plant-based enzymes with every meal. These enzymes assist the body in breaking down and assimilating the nutrients in the foods you eat regardless of pH. It was mentioned at the beginning of this chapter that if the blood becomes too acidic, it leads to an environment in the digestive tract that is less than advantageous. This leads to additional stress on the pancreas, liver and gallbladder, which in turn leads to metabolic enzyme deficiencies. These conditions can often be overcome with a high potency, digestive enzyme blend.

References:

Cousens, Gabriel. *Conscious Eating*. Berkley, CA, North Atlantic Books, 2000.

Young, Robert O, and Shelley Redford Young. *The pH Miracle*. Warner Books, Inc., 2002.

Wiley, Rudolph A. Biobalance: *The Acid/Alkaline Solution to the Food-Mood-Health Puzzle*. Essential Science Publications, 1988.

Baroody, Dr. Theodore A. *Alkalize or Die*. Waynesville, NC, Holographic Health Press, 1991.

Chapter Nine
The Microflora (Bacteria) Connection

Bacteria are simple single-celled organisms that are found everywhere on earth. They are so prevalent that it is believed they are likely the most numerous types of organisms on the planet. Like plant cells, they have both a cell membrane and a cell wall. Though most bacteria are not harmful, we often think first of those bacteria that are, such as *Streptococcus* that cause strep throat, *Staphylococcus*, the cause of "staph" infections, or *E. coli*, which is a common cause of food poisoning. In nature and science however, bacteria contribute to more good than bad. For example, bacteria has been used in the production of certain fermented dairy products such as cheese and yogurt for centuries. More recently, other types of bacteria have been used by science in the biotechnology field to produce useful medicines such as insulin for diabetics.

Within our body and in particular within our colon, bacteria play a vital role in overall health. The condition and function of the gastrointestinal tract with its trillions of bacteria and yeast (or microflora) is essential to our well-being.

By some estimates there are over 500 species of bacteria in our intestinal tract alone. That represents between 30% and 50% of the total weight of the contents of the intestinal tract.

A symbiotic relationship exists with this microflora; the beneficial microflora protects against the overgrowth of pathogenic organisms, assists in the digestion of fiber and lactose, while producing enzymes and manufacturing B vitamins. We in turn, give these friendly bacteria a place to live and food to prosper.

If the population of good bacteria diminishes to the point where many of these beneficial actions cease, opportunistic "bad" bacteria can take its place and wreak havoc on the body. One such opportunistic yeast is called *Candida albicans.* When the opportunity arises, this yeast that normally resides in the colon in relatively small numbers can overgrow, becoming a fungus. In the overgrowth state it can actually form "roots" that burrow through the intestinal lining and strip the colon of all of its beneficial microflora. This can lead to Irritable Bowel Syndrome (IBS), chronic yeast infections, skin problems, immune disorders and numerous digestive issues.

The cause of this imbalance of bacteria is most often the administration of antibiotics, which decreases metabolic activities within the colonies. Antibiotics do not discriminate; in their presence all bacteria die, the good with the bad. It has also been noted that birth control pills and cortisone may increase the possibility of a *Candida* overgrowth. To maintain and/or re-establish a healthy balance, probiotics may be the most natural, safe and commonsense approach. A probiotic is the exact opposite of an antibiotic; they are made up of beneficial bacteria and can re-colonize the digestive tract to re-establish the balance lost.

You may ask yourself at this point why would a book about enzymes have a chapter on bacteria or microflora. Well there are actually several connections between bacteria and enzymes:

1) Like enzymes, microflora are essential for good digestion.
2) Microflora produce enzymes (cellulase, lactase, protease, and amylase) to complete digestion and are synergistic with enzymes in supporting the immune system.
3) Like enzymes, microflora are instrumental in protecting the host from invading bacteria and viruses.
4) Probiotics are a great adjunct to supplemental enzymes since enzymes taken as a supplement eventually "wear out." Probiotics live in the gut, colonize and create more enzymes.

What to Look For in a Good Probiotic

When it comes to picking a probiotic there are many choices. What often makes this confusing is the number of different strains available, the process of manufacturing and the way companies label their products. Today you can find products containing one strain of probiotic and others with 20 strains. The most common are *Lactobacillus acidophilus* and *L. bifidus*; these are supported by more research and have been in use longer than any of the other choices. The problem with these strains is that they are not very hardy; they die rather easily from exposure to heat, acid and air.

Many companies that sell products with these microflora as the main ingredients will often state that the number of CFU's (colony forming units) contained in the capsule are "at time of manufacture." This is basically telling the customer that by the time they consume the product there is no telling exactly how many of these sensitive bacteria will be alive, which is one of the reasons that so much scrutiny has been placed on these products. Often fewer than half of the probiotic products tested will meet their label claim. For this reason I always recommend that you look for a probiotic that guarantees potency at the *time of consumption.*

You should look to a supplement that has a "delivery system" that ensures the friendly bacteria will survive the trip. Some bacteria strains have been shown to survive the acid environment of the stomach such as *Bacillus subtillus* and F-19, but most cannot. For the strains that have difficulty surviving, it is best to deliver them in an enteric-coated capsule that will protect the sensitive bacteria from the stomach acid, delivering it safely to the intestinal tract.

Some of the more common strains you will find (in addition to *L. acidophilus* and *L. bifidus*) include *Lactobacillus casei, Lactobacillus bulgaris, Bacillus subtillus, Lactobacillus plantarum, Lactobacillus rhamnosus, Lactobacillus salivarius, Lactobacillus longum, Lactobacillus lactis* and F-19. Obviously, unless you have studied probiotics and bacteria these strains may not mean anything you. The following is a listing of the eight strains that I believe are the most noteworthy, along with some of the studies that suggest their benefit.

LACTOBACILLUS ACIDOPHILUS
"*Lactobacillus acidophilus* helps to control diarrhea and to reduce bad cholesterol by converting it to coprostanol and enabling its elimination. It helps in the reduction of lactose intolerance and in the control of *Candida* overgrowth, it protects the gastrointestinal tract and it strengthens the immune system."[1]
"For example, the breakdown of food by *L. acidophilus* leads

to production of lactic acid, hydrogen peroxide, and other byproducts that make the environment hostile for undesired organisms. *L. acidophilus* also produces lactase, the enzyme that breaks down milk sugar (lactose) into simple sugars. People who are lactose intolerant do not produce this enzyme. For this reason, *L. acidophilus* supplements may be beneficial for these individuals."[2]

LACTOBACILLUS CASEI

"*Lactobacillus casei* is effective in the treatment of intestinal infections. It inhibits tumors in mice and increases immunity in humans against bacterial and viral infections. *L. casei* encourages proper gastrointestinal function and elimination and maintains the balance of the gastrointestinal terrain."[3]

"Oral administration of *Lactobacillus casei* has been found to enhance innate immunity by stimulating the activity of... NK cells. The ability to switch mucosal immune responses... with probiotic bacteria provides a strategy for treatment of allergic disorders."[4]

LACTOBACILLUS BULGARICUS

"*Lactobacillus bulgaricus* does not colonize in the intestinal tract, is fast growing, and produces lactic acid; it thus promotes the growth of beneficial bacteria that establish a balanced gastrointestinal tract environment. It contributes to digestion, lactose tolerance, the reduction of cholesterol, and the control of intestinal infections and it also enhances immunity."[5]

"Beneficial *L. bulgaricus* colonies form a hostile environment for pathogenic (disease-causing) germs and play a major detoxification role in removing potentially harmful germs that travel through the G. I. tract. This cleansing activity also helps sweep metabolic waste and chemical toxins from the body. *L. bulgaricus* helps in correcting the condition of either constipation or diarrhea by significantly influencing the peristaltic action of the G.I. tract. Chronic, persistent diarrhea is less common in infants fed yogurt containing *L. bulgaricus* compared to milk fed infants."[6]

LACTOBACILLUS PLANTARUM
"*Lactobacillus plantarum* produces lactic acid, inhibits the growth of gastrointestinal tract pathogens, and prevents flatulence. One strain of the *L. plantarum* species has been tested clinically for its effect on irritable bowel syndrome (IBS). In both studies, subjects showed a decrease in IBS symptoms and reduced pain."[7]

"*L. plantarum* appears to help preserve nutrients such as omega-3 fatty acids, but also increase their content. *L. plantarum* has also demonstrated the ability to reduce and eliminate potentially pathogenic microorganisms both in vitro and in vivo."[8]

"In a 4-week, double-blind, placebo-controlled trial of 60 individuals with IBS, probiotics treatment with *L. plantarum* reduced intestinal gas significantly."[9]

LACTOBACILLUS RHAMNOSUS
"*Lactobacillus rhamnosus* provides mucosal support by adhering to the mucosal membrane, inhibiting fungal or bacterial vaginal infections, and preventing infection. *L. rhamnosus GG*...reduce the incidence of or lessen the severity of antibiotic-associated diarrhea."[10]

"According to a recent study in 10 healthy adults, cellular immune response to intestinal microorganisms was enhanced following intake of *L. rhamnosus GG* for 5 weeks. This increased the response of peripheral T lymphocytes to intestinal bacteria and enhanced an inflammatory response by increasing the secretion of suppressive cytokines and decreasing secretion of pro-inflammatory cytokines."[11]

"Administration of *L. rhamnosus GG* to pregnant mother weeks prior to delivery and to their newborn babies through 6 months of age led to a 50% decrease in the infants' incidence of recurring atopic eczema."[12]

LACTOBACILLUS SALIVARIUS

"*Lactobacillus salivarius* prevents flatulence and inhibits intestinal putrefaction and the development of undesirable bacteria in the mouth and intestines. It is antibiotic resistant and therefore helps to prevent antibiotic-induced diarrhea." [13]

"*L. salivarius* is classified as a facultative bacterium, which means that it can survive and grow in both anaerobic (without oxygen) and aerobic (with oxygen) environments, although its main effects take place in anaerobic conditions. One unique benefit of *L. salivarius* is its ability to help break down undigested protein and disengage the toxins produced by protein putrefactions. In one study, *L. salivarius* was able to produce a high amount of lactic acid and completely inhibit the growth of *H. pylori* in a mixed culture. *L. salivarius* was found to be a potentially effective probiotics against *H. pylori*."[14]

BACILLUS SUBTILLIS

This bacterium is the source of nattokinase, the enzyme

discovered to support cardiovascular health and balance blood viscosity. It is one of the most studied strains in the world and has been shown to normalize intestinal microflora balance.

"*B. subtilis* is a microorganism also able to secrete large amounts of extracellular enzymes. The most abundant of these are: a-amylase, arabinase, cellulase, dextranase, levansucrase, maltase, alkaline protease, neutral protease, B. glucanase, DNase and various enzymes with N. acetylmuramidase activity."

"Members of the genus *Bacillus*, and in particular *B. subtilis*, are widely distributed in nature, particularly in soil from where they spread anywhere, and can be isolated from a wide variety of sources. *B. subtilis* is able to grow and exert a positive metabolic effect in different habitats, including soil and the gastrointestinal tract of animals and man."

"*B. subtilis* can thus be considered...a truly indigenous component of the normal flora, probably located in areas of the intestinal tract where a residual amount of oxygen is available."

"*Bacillus subtilis* is one of the most important anti-diarrhea microorganisms used in human medicine for the therapy and prophylaxis of intestinal disorders essentially produced by antibiotic therapy."

"*Bacillus subtilis*...improved the growth and viability of *Lactobacilli*...enhanced the growth and viability of *L. reuteri*... multiple catalases having differential susceptibility to the inhibitor were produced."[15]

LACTOBACILLUS F19

F19 is a member of the *Lactobacillus acidophilus paracasei* species. It is well tolerated by infants, adults, and the elderly, adheres to the colon and persists in the GI tract. It has also been shown to increase lactic acid bacteria in the GI-microflora after 8 weeks of consumption.

"*Lactobacillus* F19 is an emerging probiotic with good technological characteristics. It now has a proven ability to survive gastric transit and to persist in the colonic environment of humans."

"*Lactobacillus* F19 is indigenous to the intestinal tract of a portion of the population in Finland and Sweden."

"*Lactobacillus* F19 is well-tolerated by infants, adults, and the elderly. Adheres to the colon. Persists in the G.I. tract and, in some cases, could even be isolated 8 weeks after intake."

"*Lactobacillus* F19 increases other lactic acid bacteria in the GI-microflora."[16]

A Word about Protease and Probiotics

There has been some recent concern from a small few who suggest that protease should never be taken with probiotics. These individuals/companies have been promoting the idea that the enzyme protease "kills" both the probiotics consumed orally and the microflora that are already a part of the intestinal ecosystem. Though they may sincerely believe this to be true, it simply has no basis in fact. This assumption is based on the simplistic understanding that protease breaks down protein. Though this is true, it is an incomplete scenario.

Plant-based proteases (from aspergillus) have the ability to break down a fairly wide range of proteins in a reasonably wide pH. However, plant-based proteases do not break down all proteins. In order for these proteases to break proteins down, three criteria must first be met. First, they have to be attached to a dead organism; second, damaged; or third, attached to antibodies. These are the only proteins that such proteases have the ability to break down. If these plant-based proteins did indeed break down all proteins, then all red blood cells, white blood cells, muscle tissues, stomach lining, arteries, other enzymes and practically every other tissue of the body would be in danger!

Another point to consider is the content of the probiotics themselves. They are bacterial organisms, full of enzymes including proteases. Why don't the proteases in the probiotic "kill" the cell? The answer is obvious: the protease in bacteria is specific and does not digest the living protein found in the bacteria itself. (See Appendix for more information)

References

1 Rosell JM, Brochu E, Vezina C, et al., Rosell Institute Inc, unpublished data, 1997. Harmsen HJ, Wildeboer-Veloo AC, Raangs GC, et al. "Analysis of intestinal flora development in breast-fed and formula-fed infants by using molecular identification and detection methods." *J Pediatr Gastroenterol Nutr.* 2000;30:61-67.

2 Alvarez-Olmos MI, Oberhelman RA. "Probiotic agents and infectious diseases: a modern perspective on a traditional therapy." Clin Infect Dis. 2001 Jun 1;32 (11):1567-76. Cunningham-Rundles S, Ahrne S, Bengmark S, et al. "Probiotics and immune response." *Am J Gastroenterol.* 2000 Jan;95 (1 Suppl):S22-5.

3 The Burton Goldberg Group. Alternative Medicine, The Definitive Guide. Fife, WA: Future Medicine Publishing, Inc; 1995. Institute for Functional Medicine, Inc. *Clinical Nutrition: A Functional Approach.* Gig Harbor, WA: The Institute for Functional Medicine; 1999.

4 Matsuzaki T., Chin J. "Modulating immune responses with probiotic bacteria." *Immunology and Cell Biology*, Volume 78, Number 1, February 2000, pp. 67-73(7).

5 Institute for Functional Medicine, Inc. "Clinical Nutrition: A Functional Approach." Gig Harbor, WA: The Institute for Functional Medicine; 1999. Jarnason I, Williams P, Smethurst P, et al. "Effect of non-steroidal anti-inflammatory drugs and prostaglandins on the permeability of the human small intestine." Gut. 1986;27:1292-1297.

6 Bergogne-Berezin E. "Treatment and prevention of antibiotic associated diarrhea." *Int J Antimicrob Agents.* 2000 Dec;16(4):521-6. Bryant, M. "The shift to probiotics." *The Journal of Alternative Medicine.* 2: 6-9. 1986. Tejada-Simon, M. "Proinflammatory cytokine and nitric oxide induction in murine macrophages by cell wall and cytoplasmic extracts of lactic acid bacteria." Department of Food Science and Human Nutrition, Michigan State University, East Lansing, MI, 1999.

7 Niedzielin K, Kordecki H, Birkenfeld B. "A controlled, double-blind, randomized study on the efficacy of *Lactobacillus plantarum* 299V in patients with irritable bowel syndrome." *Eur J Gastroenterol Hepatol.* 2001 Oct;13(10):1143-7. Nobaek S, Johansson ML, Molin G, Ahrne S, Jeppsson B. "Alteration of intestinal microflora is associated with reduction in abdominal bloating and pain in patients with irritable bowel syndrome." *Am J Gastroenterol.* 2000 May;95(5):1231-8.

8 Benmark, S. *Nutrition* 14, nos. 7-8 (July 1998): 585-94.

9 Nobaek S, Johansson ML, Molin G, et al. "Alteration of intestinal microflora is associated with reduction in abdominal bloating and pain in patients with irritable bowel syndrome." *Am J Gastroenterol.* 2000; 95:1231 – 1238.

10 Heyman, M.J. *Am. Coll. Nutr.* 19: 137s, 2000.

11 Schultz M, Linde HJ, Lehn N, et al. J. *Dairy Res.* 70: 165, 2003.

12 Kalliomaki M, Salminen S, Arvilommi H., et al. *Lancet* 357: 1076, 2001.

13 JMR, EB, CV, et al., unpublished data, 1997.

14 Aiba Y, Suzuki N, Kabir AM, Takagi A, Koga Y. "Lactic acid-mediated suppression of Helicobacter pylori by the oral administration of *Lactobacillus salivarius* as a probiotic in a gnotobiotic murine model." *Am J Gastroenterol.* 1998 Nov;93(11):2097-101. Kabir, A.M. Gut 41, no. 1 (July 1997): 49-55

15 Hosoi T, Ametani A, Kiuchi K, Kaminogawa K. "Improved growth and viability

of lactobacilli in the presence of *Bacillus subtilis* (natto), catalase, or subtilisin." *Can. J. Microbiol./Rev. Can. Microbiol.* 46(10): 892-897 (2000). Piergiorio M. "The use of Bacillus subtilis as an antidiarrhoeal microorganism." Dipartimento Farmaceutico, Universita di Parma, Viale delle Scienze, 43100 Parma.

16 *Lactobacillus Paracasei* subsp. paracasei F19: Survival, Ecology and Safety in the Human Intestinal Tract—A Survey of Feeding Studies within the PROBDEMO Project. Bennet, R, Nord CE, Matto J. "Faecal recovery and absence of side effects in children given Lactobacillus F19 or placebo," 2000. Miettinen M, Voupio-Varkila J. and Varkila K. "Production of human tumour necrosis factor alpha, interleukin-6 and interleukin-10 is induced by lactic acid bacteria." *Infection and Immunity.* 64:5403-5405. 1996. Salminen S, Laine M, von Wright A, et al. "Development of selection criteria for probiotics strains to assess their potential in functional food. A Nordic and European approach." *Bioscience Microflora*, 1996, 15: 61-67.

Chapter Ten
The Immune System and Enzyme Therapy

The immune system is a whole-body network of cells and organs which, when working as intended, defends the body against attacks from "foreign" invaders. These invaders include bacteria, viruses, parasites and fungi. There are times though, when the immune system becomes either over-active or under-active. When under-active, we become vulnerable to many different diseases; when over-active, the immune system can begin to attack vital organs, tissue and cells. This condition is called an autoimmune disease and is the reason for such ailments as lupus, rheumatoid arthritis and pancreatitis.

According to the U.S. Department of Health and Human Services, "Autoimmune diseases are a family of more than 80 chronic, and often disabling, illnesses that develop when underlying defects in the immune system lead the body to attack its own organs, tissues, and cells. While many of these diseases are rare, collectively they affect 14.7 to 23.5 million people in this country, and their prevalence is rising." To truly understand the reasons why these diseases plague us we must first have a grasp of how the immune system works.

Immunity 101

In general, there are two types of immunity: the specific (or what is sometimes called adaptive immunity) and the non-specific (or what is often called innate immunity). Innate immunity (or non-specific) consists of the barrier layers such as skin and mucous secretions. Such immunity protects us from *any* intruder. It is not specific to bacteria, viruses and the like. For example, the secretions in the nose create a barrier to dust, not just the pathogens that can make us ill. On the other hand, adaptive immunity (or specific) is a direct response to a specific immune stimulus, or what is termed as an antigen. The specific immune response involves two types of cells that are the key to this defense system: B cells and T cells. When these cells are activated, invading organisms are destroyed.

ORGANS AND CELLS OF THE IMMUNE SYSTEM

The organs of the immune system are stationed throughout the body. Known as the lymphoid organs, they are connected with the growth, development, and deployment of lymphocytes (the white blood cells that play the key role in defending the body from harmful pathogens).

Cells destined to become immune cells arise in the bone marrow from the stem cells. Some of them develop into myeloid cells, a group known for their large size and ability to devour harmful pathogens. These are also known as white blood cells and more specifically, phagocytes. Phagocytes include monocytes, macrophages, and neutrophils. Other myeloid descendants are eosinophils and basophils, which more commonly are involved in allergic responses.

The organs of the immune system are connected with one another and with other organs of the body by a network of lymphatic vessels that are similar to blood vessels.

Immune cells and foreign particles are conveyed through the lymphatic system in a clear fluid that bathes the body's tissues. Lymph nodes are small, bean shaped structures that are laced throughout the body along the lymphatic routes. Found in the neck, under the arm and groin (among other places) they contain specialized compartments where immune cells congregate, and where they can encounter antigens. Lymphoid cell precursors develop into the small white blood cells called lymphocytes (B cells and T cells).

SELF AND NON-SELF

For the immune system to work properly it must be able to distinguish between self and non-self. Every body cell carries distinctive molecules that distinguish it as part of the body. This marker is its passport, if you will, which tells the immune system it has the right to dwell within the body. Normally the body's defenses do not attack tissues that carry such a marker. Immune cells coexist peaceably with other body cells in a state known as self-tolerance.

Foreign molecules, too, carry distinctive markers, characteristic shapes (epitopes) that protrude from their surfaces. One of the remarkable things about the immune system is its ability to recognize many millions of distinctive non-self molecules, and to respond by producing antibodies that can match and counteract each of the non-self molecules.

An antigen is any substance capable of triggering an immune response. It can be a bacterium, a virus or even a portion of one of these organisms. Tissues or cells from another individual also act as antigens; that's why transplanted tissues are rejected as foreign. When a person receives a heart transplant they must additionally take drugs for the rest of their lives to suppress their immune function. If they fail to do this, the body's own defenses will reject the organ as non-self, or foreign.

THE IMMUNE CELLS

A key component of the immune system are the B cells, whose chief function is to secrete soluble substances known as antibodies. Each B cell is programmed to make only one specific antibody. When a B cell encounters its triggering antigen, it gives rise to many large plasma cells. Each plasma cell is essentially a factory for producing that one specific antibody.

Antibodies belong to a family of large protein molecules known as immunoglobulins. Scientists have identified nine chemically distinct classes of human immunoglobulins which can be categorized into four types of Immunoglobulin G (IgG), two types of Immunoglobulin A (IgA), and one type Immunoglobulin M (IgM), Immunoglobulin E (IgE) and Immunoglobulin D (IgD).

T cells contribute to the immune defenses in two major ways. Some assist in regulating the complex workings of the immune system, while others directly contact infected cells and destroy them. Chief among these T cells are "helper/inducer" T cells, which are needed to activate many immune cells, including B cells and other T cells. Another subset of regulatory T cells acts to turn off or suppress immune cells. Cytotoxic, or "killer" T cells help rid the body of cells that have been infected by viruses as well as cells that have been infected with cancer. They are also the cells responsible for the rejection of tissue and organ grafts.

THE NERVOUS SYSTEM CONNECTION

Biological links between the immune system and the central nervous system exist on several levels. Hormones and other chemicals such as neuropeptides convey messages among nerve cells. Recent research has shown that they "speak" to the cells of the immune system. In addition, networks of nerve fibers have been found to connect directly to the organs of the lymph system. What emerges from this is a closely interlocked system with a two-way flow of information. Some speculate that immune cells may function in a sensory capacity, detecting the arrival of foreign invaders and relaying chemical signals to alert the brain. The brain, in turn, may send signals that guide the traffic of cells through the organs related to the lymph system.

THE COMPLEMENT SYSTEM (ENZYMES AT WORK)

The complement system consists of a series of proteins (enzymes) that work to "complement" the work of antibodies in destroying bacteria. These metabolic enzymes (or proteins) circulate in the blood in an inactive form. The complement cascade is set off when the first complement molecule, C1, encounters an antibody bound to an antigen in an antigen-antibody complex. Each of the complement enzymes performs its specialized job, acting on the next complement molecule in line. The end product is a cylinder-like shape that punctures the cell membrane and, by allowing fluids and molecules to flow in and out, dooms the target cell.

MOUNTING AN IMMUNE RESPONSE

Microbes attempting to get into the body must first get past the skin and mucous membranes, which not only pose a physical barrier but are also rich in scavenger cells and IgA antibodies. Next, they must elude a series of non-specific defenses (cells and substances that attack all invaders regardless of the epitopes they carry). These include patrolling scavenger cells, complement proteins and various other enzymes that are part of the body's natural defenses. Infectious agents that get past the non-specific barriers must then confront specific weapons produced by the body that have been specifically tailored to suit the intruder. Whenever T cells and B cells are activated, some become "memory" cells so that the next time an individual encounters that same antigen, the immune system is primed to destroy it quickly.

Long-term immunity can be stimulated not only by infection but also by vaccines made from infectious agents that have been inactivated or, more commonly, from minute portions of the microbe. An example of this is the flu vaccine, which is comprised of inactive flu virus. When presented to the immune system, the flu virus acts as a "wanted" poster, alerting the immune system to be on the watch for live flu virus when the body is exposed.

Short-term immunity, on the other hand can be transferred passively from one individual to another via antibody-containing serum. In this way, infants are protected by antibodies they receive from their mothers primarily before birth and during breastfeeding.

Allergies

When the immune system malfunctions, it can unleash a torrent of disorders and diseases. One of the most familiar is an allergic response. Allergies such as hay fever and hives are related to the antibody known as IgE. The first time an allergy-prone person is exposed to an allergen, (such as grass pollen) the individual's B cells make large amounts of the grass pollen-specific IgE antibody. These IgE molecules attach to histamine-containing cells known as mast cells, which are plentiful in the lungs, skin, tongue, and linings of the nose and gastrointestinal tract. The next time this person encounters grass pollen, the IgE-primed mast cell releases powerful chemicals that cause the wheezing, sneezing, and other symptoms of an allergic reaction.

Auto-Immune Diseases

Sometimes the immune system's recognition apparatus breaks down, and the body begins to manufacture antibodies and T cells directed against the body's own cells and organs. Such auto-antibodies contribute to many diseases. For instance, T cells that attack pancreas cells contribute to diabetes, while an auto-antibody known as the rheumatoid factor is common in people with rheumatoid arthritis.

When normal cells turn into cancer cells, some of the antigens on their surface will change. These new or altered antigens catch the attention of immune defenders, including T cells, natural killer cells, and macrophages. According to one theory, patrolling cells of the immune system provide continuing body-wide surveillance, spying out and eliminating cells that undergo malignant transformation. Tumors will only develop when this surveillance system breaks down or is overwhelmed.

Though few agree on the causes of autoimmune diseases, there are some popular theories. One has to do with the production of immune complexes, which are the result of immune responses trying to overcome a threat. The immune complexes may then settle in tissue and joints signaling the immune system to act upon those specific areas. Others believe that inflammation on a microscopic level produces this same immune reaction, while other theories place the culprit on viruses, bacteria, toxins or heavy metals. What all of these theories have in common is the fact that they all speculate that the immune system is likely looking for something it cannot find (as in the case of virus, toxin, bacteria) or it is looking for something that does not exist or poses no real threat (as in the case of circulating immune complexes).

In all such cases however, enzymes can greatly assist

the body. All the conditions mentioned above can be aided through the oral intake of proteases, the enzyme known to break down protein. As was mentioned in Chapter Six, protease is the only enzyme that has the ability to become part of the immune system. What was not already mentioned is that unlike other immune support products, protease can be taken as a supplement for auto-immune diseases. While echinacea, beta-glucans, and zinc can be supportive, they do so by stimulating an immune function. This can be a problem however, in auto-immune diseases because the immune system is already over stimulated! Protease has the ability to help take out the suspected causes of the immune system attacking the body. Thus it is better described as an immune balancer than an immune stimulator.

In 1995, Dr. Billigmann studied the effects of protease on the herpes zoster virus and published a report that suggested enzymes work at least as well as pharmaceutical drugs in removing the virus. In the controlled study with 192 patients, he concluded that overall the enzyme preparation showed identical efficacy with the drug Acyclovir.[1] The herpes zoster virus has been successfully treated with enzyme therapy since 1968.

Heavy metals, such as lead and mercury exert their poisoning effect by binding to groups of proteins, including vital enzymes. Once they bind to a functional protein, such as an enzyme, they denature and/or inhibit it. This interaction of heavy metals to proteins can lead to degenerating diseases, nerve damage or even death. Some clinical observations have shown that when proteases are taken in large amounts, heavy metal concentrations have been significantly decreased in the blood. It is possible that the activated Alpha2-macroglobulin protease complex may have a high affinity for heavy metals, leading to their removal from the body.

Back to Basics

Not to be overlooked is the importance of freeing up systemic energy in order to balance immune function. Whenever the body is in crisis, it is imperative to reduce the amount of energy needed for digestion. This simple act is one of the primary tools that can be done to help overcome the complex issue of immune problems, whether the immune system is over-active or under-active. Remember that by some estimates the act of digestion consumes as much as 80% of our daily energy. If we give some of that back, we can offset a deficiency that may be at the root of the problem. In Chapter Five the importance of resting the digestive system was stressed and four options were given. These included 1) digestive enzymes at every meal, 2) restricting calories, 3) eating a higher ratio of raw food and 4) fasting. Any of these choices will assist healing, yet together they can offer help often not realized with other healing modalities.

References
1 Billigmann VP. "Enzyme therapy—an alternative in treatment of herpes zoster. A controlled study of 192 patients" [translated from German]. *Fortschr Med*. 1995;113:43–48..

Chapter Eleven
Physical Fitness and Enzymes

The quest for optimal health and physical fitness rages on. Today there is no end to the number of supplements, articles and books dedicated to achieving this allusive goal. Gyms, health and fitness centers are springing up on every corner. People are seeking professional advice from magazines, websites and personal trainers. Yet amid all of the advice and miracle products offered, you will rarely hear anyone mentioning the important role enzymes play in achieving optimal health.

If you have ever wandered through the local health food store's sports supplement or "body building" aisle, no doubt you have noticed that there is no shortage of supplements and powders. All claim to be the "cutting edge" nutrition with the latest technology for building muscle, reducing fat, decreasing recovery time or giving you what's needed to outperform your competitor or exceed your record. Though many of these products work as advertised, a real lack of basic nutrition and "delivery" is evident. If the protein powder you're taking is poorly digested, of what real benefit is it?

Digestion is defined as the ability to convert food into a

form that can be assimilated by the body. Nutrients, whether in the form of food or supplements, can only be delivered to the cells by an enzyme or multiple enzymes. Thus, it goes without saying that proper digestion and assimilation are vital for good health and physical fitness. Often there is a tendency to fill the body with vitamins, proteins and specialty supplements that lack the necessary enzymes to deliver these essential nutrients in their proper amounts to their proper places.

The reason it is often recommended to take large amounts of supplements has little to do with the body's need for such high quantities. Rather it has to do with the body's inability to assimilate such supplements. Therefore to ensure at least some delivery, it is often recommended to take enormous amounts of specific nutrients to better our chances of delivery. It would be impossible to consume these quantities in a healthy diet of organic foods. Yet we seem perfectly content to flood the body with pills and capsules.

The solution is simple! Add a high-quality digestive enzyme to all meals and supplements consumed so that proper assimilation can occur. This crucial step can make a dramatic improvement on individuals, whether they are training for a three kilometer run or a triathlon.

As an example, it is common among those who take a lot of protein supplements to complain of digestive issues. By introducing an enzyme blend that is high in protease (to break down protein) this complaint will never be a problem again. If the digestive complaint ends, imagine the benefit to the body and the increased uptake of the protein consumed. This same premise holds true for all supplements. To better benefit from any supplement, a digestive enzyme should be consumed with each meal. Since most supplements are recommended to be taken with food, the enzymes will help deliver the nutrients in both the meal and the supplement. However, what if the manufacturer recommends the supplement be taken on an empty stomach? Once again, a broad enzyme blend is recommended for optimum delivery. A new product has recently been developed for use on an empty stomach called Enhance. Manufactured by Enzymedica, this product contains the unique blend of enzymes and Bioperine® which has been shown to increase the uptake of specific vitamins and minerals. Bioperine is a standardized extract (95% piperine, a phytonutrient and the active ingredient from the fruit of *Piper nigrum L.*, or black pepper). It is lipid soluble, so therefore it alters the structure of the cell membranes, thereby augmenting the permeability of the intestinal cell membranes, leading to greater absorption of nutrients. Also regarded as a thermo-nutrient, it can stimulate thermogenesis boosting available energy for nutrient absorption by increasing the activity of an enzyme, which breaks down ATP (adenosine triphosphate). Piperine may also enhance blood flow to the gastrointestinal tract, thereby increasing the levels of enzymes involved in the transport of nutrients into the intestinal cells.

Bioperine has also been the focus of several clinical studies with healthy volunteers in the U.S, which measured the absorption of three distinct categories of products. The categories evaluated with and without Bioperine were fat-soluble *(beta-carotene)*, water-soluble (vitamin B6) and a mineral (selenium, in the form of selenomethionine). Gastrointestinal absorption of all the studied nutrients, as measured by amounts present in the blood, increased dramatically when administered with Bioperine as compared to the control group receiving the nutrient alone. Selenium levels increased by 30%, beta-carotene increased by 60%, and the vitamin B6 increase was slightly higher than beta-carotene. All studies used Bioperine in the amount of 5 mg per dose. (Enhance contains 12 mgs.)

Increase Performance

Athletes are always looking for ways to perform at peak levels; enzymes can assist in doing that. In a recent study, runners were given protease supplements or a placebo four times a day for three days, 24 hours before and 48 hours after running. Then the participants were evaluated for mood, muscle soreness and pain threshold 24 and 48 hours after running 30 minutes at about 7 miles per hour. The result was less soreness, improved pain threshold and no mood change with those taking the protease supplements over the group who took the placebo.

Throughout the world, this is not a top news story. According to some sports medicine reports, there is scarcely a top athlete in Germany who is not familiar with enzyme therapy. Additionally, in Australia, athletes, runners, wrestlers, boxers, handball players and skiers are provided with enzyme capsules as a precaution to aid in rapid recovery after injury.

The enzyme-support mechanism has to do with circulation and oxygen availability. Proteolytic enzymes help the blood to circulate more effectively throughout the body by reducing fibrin (a protein produced in the blood), which decreases circulation and reduces oxygen uptake. The more efficiently the blood travels through the many miles of veins and arteries, the more long-term energy an athlete has to feed the muscles the oxygen required during hard physical labor or exercise.

ATP *(adenosine triphosphate)* is a compound that provides energy to every cell in the body. In fact it is often stated that it is the most widely distributed high-energy compound within the human body. ATP is produced in the cells by means of the Krebs cycle (See figure 11-1 below). The Krebs (or the citric acid) cycle is the common pathway to completely oxidize fuel molecules in ten steps of reactions that yield energy and CO_2. In order for each of the ten steps to occur, an enzyme must be present.

For example, the enzyme Citrate-synthase is needed in step one to hydrolyze citryl CoA to citrate and CoA. A deficiency in any one of these enzymes can inhibit this production. Though the specific enzymes cannot be supplied for this process, a deficiency may be prevented by supporting the body with other enzymes (such as digestive enzymes). When the need for digestive enzyme production decreases, the production of metabolic enzymes may increase.

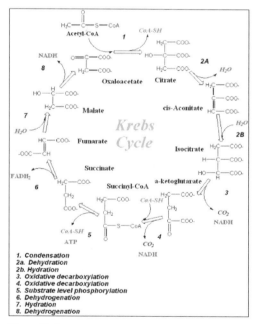

1. Condensation
2a. Dehydration
2b. Hydration
3. Oxidative decarboxylation
4. Oxidative decarboxylation
5. Substrate level phosphorylation
6. Dehydrogenation
7. Hydration
8. Dehydrogenation

Figure 11-1

Sports Injuries and Recovery

As mentioned above, in European countries, enzymes are used to enhance performance and recovery. After a sports injury, a series of metabolic processes (often known as inflammation) takes place. One of the major concerns of inflammation is capillary blood flow, so cooling, compression and rest are usually recommended as treatments.

As the smallest blood vessels in the body, capillaries are responsible for carrying oxygen and nutrients to the cells and removing waste. After an injury some of these capillaries may be damaged, making them incapable of carrying fluid to and from the damaged tissue. This leads to what is termed "walling off" of the damaged area from fibrin build-up and blockage. The result is pain, swelling, redness, heat and loss of function.

To repair the capillaries, some type of anti-inflammatory drug (aspirin, ibuprofen, etc.) is often recommended. As a result, the bruises, swelling and pain subside. The desired effect is to reduce the amount of fibrin in the damaged capillary, improve circulation and speed healing. However, this can be done very efficiently with protease (proteolytic enzymes) instead of the usual drugs. Once in the bloodstream, protease hydrolyzes (digests) the fibrin network and enhances blood flow. Additionally, these same proteases have been known to stimulate phagocytes (cells that ingest foreign particles and debris) and accelerate elimination by way of the lymphatic system.

The Athlete and pH

Strenuous exercise produces lactic acid, which is the break down of glucose and glycogen produced during a process called glycolysis. Glycolysis can happen very quickly during strenuous exercise. When it does, the formation of pyruvate exceeds the capacity of the mitochondria to accept pyruvate into the Krebs cycle. It is this excess pyruvate that is converted to lactic acid.

The increased production of this acid can slow metabolic function. As described in Chapter Eight, this can lead to an insulin imbalance and a slower metabolism, which in turn makes it more difficult to burn fat, the opposite effect of what the typical athlete is striving for. A simple solution is to eat more alkaline forming foods and consume an alkaline forming supplement (such as pH Basic by Enzymedica).

Enzymes have been described as the "energy of life" and are vital to all living organisms. To achieve the status of health and fitness you desire, add a high quality plant-based digestive supplement and a high potency protease enzyme product to your daily regimen.

References:

Bailey S.P. "Effects of Protease supplementation on muscle soreness following downhill running." Medical University of South Carolina, Charleston, SC, 1999, Chichoke, A.J., Therapeutic Uses for Enzymes, *Nutrition Science News* 7:95.

Fuller, D. "The Healing Power of Enzymes."

Gardner, M.L., *Ann Rev Nutr*, 8:329-350. 1988.

Gerbert, G., Physiologie, Stuttgart, Germany; Schattauer Klaschka, F. Oral Enzymes: new approach to cancer treatment. Munich, Germany: *Forum-Medizin*, (1966) 121.

Kleine M. Pabst H. The study of an oral enzyme therapy on an experimental erzeugre hamatome.

Kimber I, et al. *Forum of Practical and General Practitioners*, "Toxicology of protein allergenicty: prediction and characterization." 1988;27:42.

Mamadou, M., *Toxical Sci* 1999 Apr;48(2):157-62 "Oral Enzymes: Facts and Concepts."

Bioperine research:
Indian J Exp Biol 1998 Jan;36(1):46-50.

Mol Cell Biochem 1998 Dec; 189(1-2):113-8.

Nahrung 2000 Feb; 44(1):42-6.

Majeed, M., Badmaev, V., and Rajendran, R. (1996) Use of piperine to increase the bioavailability of nutritional compounds. Kawada, T. et al. (1988) *Proc. Soc. Exp. Biol. Med.* 188, 229-233.

Dora, K.A. and Clark, M.G. (1994) *Life Sci.* 55(5), 389-397.

Chapter Twelve
Choosing the Right Enzyme Supplement

When it's time to go to the health food store and buy an enzyme product, it can be a bit confusing. Often you will find that enzymes are grouped together in a way that makes little sense. The bromelain and papain products are mixed in with the digestive aids and animal source enzymes (pancreatin, trypsin and chymotrypsin). You may also find detoxifying products and system cleanses on the same shelves.

The key in choosing the appropriate enzyme supplement for you is to know what you are looking for. Each enzyme, whether of plant or animal source, serves a unique purpose. If you are looking for an enzyme formula that can help with inflammation, then you will be looking for either a bromelain, papain blend or an animal source enzyme blend. If you are trying to reduce stress digestively, you will be looking for a high potency plant-based (fungal, microbial) enzyme product. Once you have made this distinction, there are a few things to look for when picking a reputable brand.

1) Look for a company that specializes in enzymes. Often the

best place to get your brakes fixed is at a shop that specializes in brakes. This is also true of tires, transmissions, paint and so on. The reason is that this is what they know, usually better than anyone else. When buying supplements I always recommend you buy flax oil, herbs, detoxifiers, vitamins, and teas from companies that specialize in those products. Most often you will find the best products come from companies who are not trying to be all things to everyone. Though there are not a lot of companies that specialize in enzymes, the ones that do make very reputable and effective products.

2) Check the potency in the Supplement Facts. Although potency is often difficult to assess (See Chapter Seven), the best products contain high active units (not milligrams) with multiple strains in each category. This is often expressed as a blend of protease, lipase, cellulase or amylase. This blending will allow the enzymes to break down more protein, fat and carbohydrates over a longer period of time.

3) Find a product with no fillers added. Many products are formulated in a way that requires fillers to be added to either fill out the capsule or help bind the tablet together. These fillers may include Magnesium Stearate, cellulose, pectins, maltodextrin, talc and the like. The fillers differ but the result is the same: a product that is less potent per milligram and runs a greater risk of containing an allergen.

4) Find a company that tests the product to ensure that it meets the label claim. Often products are blended to meet the potency printed on the label, yet by the time they have been encapsulated or tableted, some of the activity has been lost. Reputable companies will test the finished product to make sure that what the label describes is actually in the bottle.

5) Buy enzymes in capsules (preferably vegetarian capsules) instead of tablets. Tableting is a harsh process for enzyme products since they are more susceptible to heat and friction than ordinary vitamins. In addition, the process of tableting may also require binders or fillers.

There are two companies that continue to meet all of the above requirements. The first is Enzymedica, which exclusively manufactures enzyme supplements. These are found in most health food stores. At the time of writing, Enzymedica also happens to be the most popular choice among consumers in that market. The second company is called Theramedix and they, too meet all of the above requirements but are sold exclusively through health professionals in alternative care facilities.

Though Enzymedica and Theramedix are highly recommended, there are other choices that are also satisfactory. A few of these include: Enzymatic Therapy, Renew Life, and Garden of Life.

RECOMMENDED ENZYME PRODUCT COMPANIES:
Enzymedica (carries a full line of enzyme products)
888-918-1118
www.enzymedica.com

Enzymatic Therapy (digestive, anti-inflammatory enzyme products)
800-783-2286
www.enzy.com

Renew Life (digestive, Candida enzyme products)
800-430-4778
www.renewlife.com

Garden of Life (digestive, anti-inflammatory enzyme products)
561-748-2477
www.gardenoflife.com

RECOMMENDED HEALTH PROFESSIONAL PRODUCTS:
Theramedix
866-998-4372
www.theramedix.net

Transformation Enzymes
713-266-2117
www.transformationenzymes.com

Conclusion

It has been said that the more we come to know, the more we realize how little we know.... Everything we have learned is nothing compared to what we have yet to learn. How true those words are! The reality is we know very little about life and human health. To be sure, science has made great strides but in comparison to what we have yet to learn, we are still very ignorant as to why people get sick and what it takes to make them better. Nothing shouts that louder than science's inability to manage the common cold, not to mention potentially debilitating illnesses such as cancer, heart disease, arthritis, diabetes, depression and the like.

I believe that in many ways science has become intoxicated with the concept of knowledge. Although I would not discourage anyone from pursuing such knowledge, there is much we do not know and perhaps never will. Is it possible that maybe, just maybe we are making too much of all of this?

Science has ignored the basics of health: diet and exercise. Modern medicine has become so involved in the science of disease and the relief of the symptoms of disease that they have turned a blind eye to the basics. In many ways biology and health have been written off as being too simplistic. Many prefer to go through life ignoring these essential elements of health and end up trying to compensate later by popping pills and taking pharmaceuticals. Don't be fooled by the beautiful simplicity of enzyme therapy: a balanced pH and the importance of good bacteria. The fact is though enzyme therapy is relatively easy to understand, there is still so much we have yet to learn!

PART TWO

Protocols
In Alphabetical Order

In this section, various protocols are outlined to address specific aliments. There are many books on the market that outline numerous nutritional supplements for everything you can imagine. This is an attempt to cover the most common health issues from a different angle...enzyme therapy.

The protocols do not recommend specific products. Rather they suggest a basic formula to look for and guidelines for usage. In addition to the minimum dosage suggested, look for products that are third party tested, contain no fillers and are vegan and vegetarian.

The protocols here recommended are based on eight years of practical application with thousands of individuals having tried and tested them for their effectiveness.

Anti-inflammatory Formula: To address inflammation, speed recovery, and repair tissue. Best if enteric coated.

Each serving:

Protease blend	65,000 HUT
Papain	750,000 PU
Bromelain	600 GDU(11.25 million FCCPU)

Amylase blend	4,000 DU
Lipase blend	300 FCCFIP
Catalase	50 Baker

Digestive Formula: This formula will enhance the digestion and assimilation of food while reducing the body's need to produce digestive enzymes.

Each should provide approximately:

Necessary Ingredients:

Amylase blend	12,000 DU
Protease blend	42,000 HUT
Invertase	1.75 INVU
Maltase	200 DP
Cellulase blend	200 CU
Alphagalactosidase	75 GAL
Lipase blend	500 FCCFIP
Lactase	850 LacU
Phytase	50 endo PG (or PU)
Pectinase	50 endo PG (or AJDU)

Helpful ingredients:
Xylanase
Betaglucanase
Hemmicellulase
L. acidophilus
L. bifidus

High Potency Digestive Formula: The higher potency formula will average about three times the potency of the average digestive formula. An average formula may be substituted if three times the regular dose is taken.

Each capsule should contain approximately:

Amylase blend	22,000 DU
Protease blend	80,000 HUT
Lipase blend	3,000 FCCFIP
Cellulase blend	2000 CU
Invertase	80 INVU
Lactase	900 LacU
Maltase	200 DP
Glucoamylase	50 AG
Alpha-Galactosidase	450 GAL
Phytase	50 endo-PG
Pectinase	50 AJDU
Xylanase	500 XU
Hemmicellulase	30 HCU
Beta Glucanase	25 BGU
L. acidophilus	250 Million CFU

High Lactase Formula: The high lactase formula is for individuals who have difficulty digesting foods that contain lactose (a dairy sugar), along with dairy fat and dairy protein.

Each capsule should contain approximately:

Necessary Ingredients:
Lactase 9,000 ALU
Protease blend 20,000 HUT
Lipase blend 500 FCCFIP
Helpful ingredients:
Maltase 300 DP
Cellulase blend 300 CU
Alpha-Galactosidase 50 GALU
Invertase 25 INVU
Amylase blend 7,500 DU
Glucoamylase 20 AG

Soothing Digestive Formula: Helps alleviate conditions associated with gastro- intestinal distress.

Each capsule should contain approximately:

Amylase blend	2,000 DU
Lipase blend	175 FCCFIP
Cellulase blend	400 CU
Marshmallow Root	100 mg
Gotu Kola	50mg
Papaya Leaf	100 mg
Prickly Ash Bark	50mg

This formula should not contain protease, although other herbs may also be present.

High Amylase Formula: The high amylase enzymes can be used for overcoming symptoms of allergies and for the proper digestion of carbohydrates, especially grains, raw vegetables and legumes.

Each capsule should contain approximately:

Amylase blend	22,000 DU
Glucoamylase	30 AG
Alpha-Galactosidase	1,000 GAL*
Protease blend	15,000 HUT
Lipase blend	150 FCCFIP
Cellulase blend	400 CU
Lactase	300 LacU
Maltase	300 DP
Pectinase	20 endo-PG

*1000 GAL = 2,000 AGSU

High Cellulase / Candida Formula: Used to manage yeast overgrowth.

Each capsule should contain approximately:

Cellulase blend	30,000 CU
Protease blend	100,000 HUT

Contra-indications: High amounts of cellulase should not be taken with certain timed release medications that contain cellulose.

High Lipase Formula: This formula will aid in fat digestion and metabolism as well as the health of the cardiovascular system.

Each capsule should contain approximately:

Lipase blend	2000 FCCFIP
Amylase blend	5,500 DU
Protease blend	20,000 HUT
Lactase	300 LacU

High Protease Formula: To help support immune function and assist in removing viruses, fungal forms, toxins, bacteria and heavy metals.

Each capsule should contain approximately:

Protease blend	150,000 HUT
Seaprose	4 mg (or mucolase)
Catalase	50 Baker
Serratiopeptidase	25,000 units
Nattokinase blend	400 FU

Nattokinase Formula: Nattokinase NSK-SD has been shown to have a high fibrinolytic activity (breaks down fibrin), supports cardiovascular health and decreases blood pressure.

Each capsule should contain approximately:

Necessary ingredient:

Nattokinase NSK-SD	1,000 FU

Helpful ingredients:

Amylase blend	9,000 DU
Protease blend	20,000 HUT
Minerals	85 mg
Glucoamylase	25 AGU
Lipase blend	1,000 FCCFIP
Cellulase blend	400 CU

Nutrient Enhance Formula: Helps the body benefit from supplemental vitamins, minerals and herbs.

Each capsule should contain approximately:

Bioperine	10 mgs
Amylase blend	6,000 DU
Protease blend	10,000 HUT
Maltase	100 DP
Glucoamylase	25 AG
Alpha-Galactosidase	250 GALU
Lipase blend	400 FCCFIP
Cellulase blend	500 CU
Lactase	150 ALU
Beta Glucanase	90 BGU
Xylanase	150 XU

Pectinase	40 PU
Hemicellulase	800 HCU
Invertase	79 INVU
L. acidophilus	150 million CFU

pH Balancing Formula: The blend should not exceed 8.0 on the pH scale and the capsule should be enteric coated.

Each capsule should contain approximately:

Amylase blend	2,500.DU
Cellulase blend	6,000 CU
Protease blend	1,000 HUT
Lipase blend	175 FCCFIP
Pectinase / Phytase	200 endo-Pgu

Mineral Blend: Potasium Bicarbonate, Sodium Bicarbonate, Magnesium Citrate

Herbal Blend (Hydrilla, Marshmallow, Papaya)

Probiotic Formula: Assists the body in balancing microflora.

Each capsule should contain approximately:

Probiotic Live Cells	5 billion CFU guaranteed potency
Bacillus Subtillis	No less than 3 billion
L. paracassei F-19™	No less than 1 billion**
The following blend:	No less than 1 billion**
L. acidophilus —	
L. casei —	
L. bulgaris —	
L. plantarum —	

L. rhamnosus `   _
L. salivarius   _

Seaprose / Mucolase Enzyme Formula: Mucolytic
Enzyme Formula

Each capsule should contain approximately:

Seaprose 15 mg or Mucolase 30 mg

Supporting enzymes:

Amylase blend	7,000 DU
Protease blend	20,000 HUT
Glucoamylase	25 AG
Beta-glucanase	30 BGU
Lipase blend	250 FCCFIP
Alpha-Galactosidase	50 GALU
Cellulase blend	200 CU
Xylanase	250 XU
Pectinase w/ Phytase	175 endo-PGU
Hemicellulase	30 HCU
Invertase	5 INVU

THE BASICS:

These first five recommendations address the common needs of many people. They include immune, nutrient, and energy support, with the addition of a cleansing and a rebuilding program. Not specifically mentioned in each recommendation, but not to be overlooked, is the need for a good multi-vitamin (or supplemental greens product), essential fats (flax oil or fish oil), and plenty of pure water daily.

OPTIMAL (1) DIGESTIVE AND IMMUNE SUPPORT

Enzyme deficiencies, unhealthy eating habits, exposure to environmental allergens, xenobiotics, mycotoxins and stress can all deplete and limit our body's capacity to digest, absorb and assimilate our foods completely. This, in turn, can compromise our immune system. To strengthen and optimize the digestive process and balance the immune system, the following products should be essential to your daily regimen:

Enzyme Supplementation Suggestions:

- **High Potency Digestive Enzyme Formula** with every cooked meal
- **High Protease Enzyme Formula** three times daily between meals
- **pH Balancing Formula** three times daily between meals

OPTIMAL (2) NUTRIENT SUPPORT

Many diets and foods lack the proper quantities of vitamins and minerals that are necessary in every chemical reaction that takes place in our body. Deficiencies in these vital elements can lead to a host of physical imbalances and illnesses. Supplementing the diet with a pure and balanced vitamin/mineral product can supply the body with the materials that it requires to maintain and support overall health.

Enzyme Supplementation Suggestions:

- **pH Balancing Formula** at bedtime
- **High Protease Enzyme Formula** three times daily on an empty stomach
- **High Potency Digestive Enzyme Formula** with every cooked meal
- **Nutrient Enhance Formula** with all supplements consumed on empty stomach

OPTIMAL (3) ENERGY AND ENDOCRINE SUPPORT

Adults and individuals with hormonal imbalances such as CFS (Chronic Fatigue Syndrome) and Fibromyalgia need extra support to maintain optimum levels of health. Optimizing the endocrine levels in the body can have the benefits of improving memory and concentration, rejuvenating the immune system, increasing overall energy, and feeding and fortifying the hypothalamus.

Enzyme Supplementation Suggestions:

- **High Potency Digestive Enzyme Formula** with every cooked meal
- **High Protease Enzyme Formula** three times daily on an empty stomach
- **Anti-inflammatory Enzyme Formula** three times daily on an empty stomach
- **pH Balancing Formula at bedtime**

CLEANSE AND FORTIFY PROGRAM

Individuals who have very toxic bodies need to cleanse

and re-colonize their entire digestive system with healthy bacteria. Toxins, certain drugs, fungus and chemicals in the body can interfere with the body's abilities to properly digest, absorb and assimilate nutrients that are essential to sustaining life and health. Vital organisms that live in the flora of the digestive tract are frequently destroyed by these toxins, which in turn can ultimately deplete the immune system. Toxicity of the bowel can lead to an assault on the immune system. This can place an added burden on the kidneys, lungs, liver and skin. In addition to cleansing, steps to fortify the system must also be taken to secure a healthy body and viable immune system.

Enzyme Supplementation Suggestions:

- **High Potency Digestive Enzyme Formula**
  with every meal
- **High Protease Enzyme Formula** three times daily
- **pH Balancing Formula** three times daily
- **High Cellulase/Candida Formula** at bedtime
  (to assist regular bowel movement)

*Possibly add a therapeutic enzyme for a specific problem i.e. Histamine = High Amylase / Allergy Formula, Blood Clot = Nattokinase Formula, Mucus in lungs = Seaprose / Mucolase Enzyme Formula

REBUILD PROGRAM

The Rebuild Program is designed to aid in the overall function and maintenance of the digestive system, help restore the immune system, and provide the energy, support and stamina needed to promote both healing and restoration necessary for obtaining restful sleep. This program is designed for the individual whose system has been compromised by stress, disease, trauma, or injury.

Enzyme Supplementation Suggestions:

- **High Potency Digestive Enzyme Formula** with every meal
- **Soothing Digestive Formula** with every meal
- **Anti-inflammatory Enzyme Formula** three times daily
- **High Protease Enzyme Formula** three times daily
- **High Amylase / Allergy Formula** three times daily (increase glucose levels for energy)

PROTOCOLS FOR THE BODY'S IMBALANCES

Though it may seem redundant a digestive enzyme blend with meals is mentioned in every recommendation. This is absolutely essential to ensure optimal support when taking enzymes therapeutically.

ACNE

Acne is a chronic skin disorder caused by a variety of different contributing factors. These may include but are not limited to: hormonal imbalances; stress; inflammation of the hair follicles and the sebaceous glands; a vitamin and/or mineral deficiency; a reaction to antibiotics; an overgrowth of Candidiasis; and an inability to break down sugar and trans-fatty acids. This program will cleanse the blood, fortify the endocrine system, and aid in the assimilation of sugars and fats,

Enzyme Supplementation Suggestions:

- **High Potency Digestive Enzyme Formula** with every cooked meal
- **High Amylase / Allergy Formula** three times daily between meals
- **High Lipase Enzyme Formula** three times daily between meals

Optional: add a **High Protease Enzyme Formula** between meals

ADRENAL INSUFFICIENCY

Adrenal insufficiency may be caused by inadequate hormone production. The adrenal (suprarenal) glands are the primary organ system for handling the negative effects of stress. Symptoms of adrenal insufficiency include fatigue, muscular weakness, muscle and joint pain, gastrointestinal indications, allergic hypersensitivities, hypo- or hypertension, low blood sugar and food cravings.

Enzyme Supplementation Suggestions:

- **High Potency Digestive Enzyme Formula** with every cooked meal
- **High Protease Enzyme Formula** three times daily between meals
- **pH Balancing Formula** before bed

AGE SPOTS

Age Spots are often called Liver Spots. Most often seen as brown, flat spots on the skin, they are especially found on the hands, arms and face. As the body ages they can found almost anywhere on the skin. Age spots are believed to be the result of free radical damage within the skin cells. Once thought to be merely harmless, they serve as a signal of free radical damage within the body. The spots tend to disappear, however when the liver and blood are cleansed. Causes of this condition include excessive sun exposure, compromised liver function, the ingestion of rancid oils and a relatively poor diet, and lack of exercise.

Recommendations include liver detoxification, the addition of enzymes to a well-balanced diet, and supplemental nutritional products to improve one's overall digestion and absorption of nutrients. In addition, avoiding excess exposure to the sun and nutritional support of the hormonal and immune systems would be highly favorable.

Enzyme Supplementation Suggestions:

- **High Potency Digestive Enzyme Formula** with every cooked meal
- **High Protease Enzyme Formula** three times daily between meals
- **High Lipase Enzyme Formula** three times daily between meals

AGING

Although there are no miracle cures to halt the aging process, there are healthy steps someone can take to increase their life span and slow their aging process. These steps include eating a well-balanced diet high in vitamins, minerals, antioxidants and nutrients, regular exercise, stress management, the addition of enzymes and supplemental nutritional products to optimize digestion and absorption of nutrients, and nutritional support of the hormonal and immune systems.

Enzyme Supplementation Suggestions:

- **High Potency Digestive Enzyme Formula** with every meal
- **High Protease Enzyme Formula** three times daily between meals
- **pH Balancing Formula** three times daily between meals
- **Anti-inflammatory Enzyme Formula** as needed for pain or inflammation

ALCOHOLISM (Recovery)

Alcoholism is one of the most critical health problems facing today's health practitioners. Over eighteen million people in the USA (5% of the population) are alcoholics. The physical consequences of long-term alcohol consumption include (but are not limited to) adrenal exhaustion, brain degeneration, liver degeneration and cirrhosis of the liver, hypoglycemia, metabolic damage to every cell, osteoporosis, pancreatitis, and vitamin and mineral deficiencies.

Recommendations for the recovering alcoholic include liver and blood detoxification, eating a well balanced diet with the addition of enzymes, minerals, antioxidants and supplemental nutritional products to strengthen overall digestion and absorption of nutrients, and adding probiotic supplements to restore the flora of the GI tract.

Enzyme Supplementation Suggestions:

- **High Potency Digestive Enzyme Formula** with every meal
- **High Protease Enzyme Formula** three times daily between meals
- **High Amylase / Allergy Formula** as needed to raise energy level
- **pH Balancing Formula** two times per day between meals

ALOPECIA [See HAIR LOSS]

ALLERGIES (Food)

It is estimated that nearly 60% of the American population, including 5% of young children, suffer from food allergies. Symptoms of food allergies can affect every part of the body; allergic reactions can produce mildly uncomfortable symptoms or severe illnesses. The diseases and symptoms commonly associated with food allergies can include asthma, indigestion, diarrhea, fatigue, canker sores, celiac disease, IBS, hyperactivity, ear infections, migraines, bedwetting, acne, eczema, or edema.

Many experts advocate rotation diets to identify specific food allergens. Most physicians agree that the simplest and most effective approach to treating food allergies is through the avoidance of eating allergenic foods. Since food allergies can compromise the integrity of the digestive system and exhaust the immune system, the nutritional support of these two major systems is essential. In addition, the soothing of the mucosal linings and the removal of toxins would also be suggested.

Enzyme Supplementation Suggestions:

- **High Potency Digestive Enzyme Formula** with every meal
- **Soothing Digestive Formula** as needed after meals
- **High Protease Enzyme Formula** three times daily between meals
- **Seaprose / Mucolase Enzyme Formula** or **High Amylase / Allergy Formula** as needed

ANEMIA

Anemia is a lower than normal number of red blood cells in the blood. It refers to a condition in which the concentration of the oxygen-carrying pigment (hemoglobin) in the blood is below normal. Anemia can be caused by excessive blood loss, excessive red blood cell destruction, iron deficiency, and/or deficient red blood cell production. Symptoms of anemia include weakness, fatigue, pallor, headaches, heart palpitations, shortness of breath, bruising easily, nosebleeds, bleeding gums, or frequent infections.

A blood test can include a low volume of blood, a low level of total red blood cells or abnormal size or shape of red blood cells. Identifying the underlying cause of anemia through a complete diagnostic evaluation is essential. Moreover, supporting the digestive and immune systems, and restoring energy and vital minerals and vitamins are essential.

Enzyme Supplementation Suggestions:

- **High Potency Digestive Enzyme Formula** with every meal
- **pH Balancing Formula** three times daily
- **High Protease Enzyme Formula** three times daily

ALZHEIMER'S DISEASE

Alzheimer's disease is a degenerative brain disorder suffered by 4.5 million Americans, affecting those parts of the brain that control thought, memory and language. Alzheimer's disease is a progressive mental deterioration. It is characterized by an inability to carry out activities of daily life, with a loss of cognitive and memory functions. Extensive research studies indicate that the causes of Alzheimer's disease can include genetic factors, age, environmental factors, chronic exposure to aluminum and/or silicon, and increased oxidative damage due to long-term toxic exposure.

Recommendations include a healthy, well-balanced diet and lifestyle, avoidance of products that contain mercury and aluminum (which include antiperspirants, antacids, baking powder and aluminum cookware). Nutritional supplementation with high-potency vitamins, minerals, antioxidants, in addition to magnesium and potassium products are highly suggested.

Enzyme Supplementation Suggestions:

- **High Potency Digestive Enzyme Formula** with every meal
- **High Protease Enzyme Formula** three times daily
- **Nattokinase Formula** three times daily between meals
- **pH Balancing Formula** three times daily

ANXIETY

Anxiety is an emotional state ranging from unease to intense fear. It is estimated that over four million Americans suffer from anxiety. Anxiety becomes symptomatic when it starts to inhibit thoughts and feelings and disrupt normal activities of daily life. Some of the most common symptoms include back pain, heart palpitations, excessive sweating, headaches, an inability to take in enough air, a tendency either to hyperventilate or sigh repeatedly, dizziness, digestive disturbances, and muscle tightness. Long-term anxiety can deplete the body of nutrients and minerals and contribute to glandular and hormonal imbalances.

Recommendations include avoiding caffeine, sugar, food allergens and alcohol. In addition, eating regular healthy meals in a relaxed atmosphere, incorporating relaxation, breathing exercises and moderate physical exercise into one's daily routine, and eliminating or reducing the sources of stress are also recommended practices.

Enzyme Supplementation Suggestions:

- **High Potency Digestive Enzyme Formula** with every meal
- **High Amylase / Allergy Formula** when symptoms arise
- **High Protease Enzyme Formula** three times daily between meals
- **pH Balancing Formula** three times daily between meals

ARTHRITIS

Arthritis is characterized by inflammation, pain, swelling stiffness and/or redness of the joints. The two primary types of arthritis include osteoarthritis and rheumatoid arthritis. The symptoms can vary from slight discomfort to complete debilitation.

OSTEOARTHRITIS (Degenerative Arthritis) is the most common type of arthritis resulting from wear and tear on the joints. This type most commonly afflicts the elderly.

RHEUMATOID ARTHRITIS is an autoimmune disorder. It is the most severe type of inflammatory joint disease. The body's immune system acts against and damages joints and the surrounding soft tissue. As a result, the joints in the hands, feet and/or arms become extremely painful, stiff and deformed.

Enzyme Supplementation Suggestions:

- **High Potency Digestive Enzyme Formula**

with every meal
- **Anti-inflammatory Enzyme Formula** three times daily between meals (with an additional amount used if needed)
- **pH Balancing Formula** three times daily between meals

Optional: **High Protease Enzyme Formula** three times daily between meals

ASTHMA [See also HAY FEVER]

The narrowing of the bronchioles causes asthma. Asthma is the reoccurrence of attacks of breathlessness, accompanied by wheezing when exhaling. Frequently beginning in childhood asthma often becomes less severe as the person ages.

The two main types of asthma are extrinsic and intrinsic. Extrinsic asthma is an allergy to something inhaled; intrinsic asthma tends to develop later in life and often follows a respiratory tract infection. Emotional factors, such as stress or anxiety, may precipitate attacks.

Heredity is a major factor in the development of extrinsic asthma. Recommendations include providing support of the digestive and immune systems, avoiding airborne and food allergens, cleansing the system through elimination diets, managing stress, and helping to heal irritated mucosal linings.

Enzyme Supplementation Suggestions:

- **High Potency Digestive Enzyme Formula** or High Amylase Digestion Formula with every meal

- **High Amylase / Allergy Formula** three times daily
- **High Protease Enzyme Formula** three times daily between meals

ATHLETE'S FOOT

Athlete's foot is a skin disease caused by a fungal infection, usually occurring between the toes. It is particularly prevalent and highly contagious in damp, warm places such as locker rooms, swimming pools, showers, and gyms. Symptoms include burning, itching, inflammation, blisters and scaling of the skin.

Recommendations include eating a well-balanced diet, supporting both the immune and digestive systems, and reducing intake of foods that are high in sugar.

Enzyme Supplementation Suggestions:

- **High Potency Digestive Enzyme Formula** with every meal
- **High Cellulase / Candida Formula** three times daily
- **High Protease Enzyme Formula** three times daily

AUTISM

Autism is an ailment affecting an estimated one out of every five hundred children. Autistic children demonstrate deficits in social interactions, verbal and non-verbal communication, and/or repetitive behaviors or interests.

Speech development is often delayed or absent. Some children exhibit nonsense babbling and may be unresponsive to love and affection. Many of these children have learning disabilities, are withdrawn, or have unusual responses to sensory experiences. Behavior can range from total silence to periods of hyperactivity that may include self-abuse. The cause of autism is not known. Recommendations include support of the digestive and immune systems, and ruling out food allergies, chemical sensitivities and heavy metal toxicity.

Enzyme Supplementation Suggestions:

- **High Amylase Digestion Formula** with every meal (low and slow)
- **Soothing Digestive Formula** as needed
- **High Protease Enzyme Formula** three times daily between meals (low and slow)
- **pH Balancing Formul**a three times daily

For further autism protocols / support, www.enzymestuff.com

BACKACHE

Backaches can involve disorders of the bones, muscles, nerves, joints, ligaments and tendons of the spine. Common causes can include, but are not limited to poor posture, stress, arthritis, bone disease, kidney and bladder problems, scoliosis and slipped disc.

Recommendations are dependent upon a thorough evaluation of the possible causative factors.

Enzyme Supplementation Suggestions:

- **High Potency Digestive Enzyme Formula**
  with every meal
- **Anti-inflammatory Enzyme Formula** three
  times daily (more may be used as needed)
- **pH Balancing Formula** three times daily
  between meals

Optional: Add **High Protease Enzyme Formula** to
Anti-inflammatory Enzyme Formula

BAD BREATH [See HALITOSIS]

BEDSORES

Bedsores (decubitus ulcers, pressure sores or pressure ulcers) are areas of damaged skin and tissues that develop when sustained pressure, usually from a bed or wheelchair, cuts off blood circulation. Without adequate blood flow, the affected tissue dies. They are most commonly found on the hips, shoulder blades, buttocks, sacrum, and heels of comatose patients, the bedridden, and paraplegic patients. In addition to frequent turning and providing a clean environment, support of the digestive and immune systems, and controlling any inflammation is essential.

Enzyme Supplementation Suggestions:

- **High Potency Digestive Enzyme Formula**
  with every meal
- **High Protease Enzyme Formula** three times
  daily between meals

* **Nattokinase Formula** three times daily between meals

Optional: **Anti-inflammatory Enzyme Formula** given for excess pain or inflammation

BLADDER

The bladder is the hollow, muscular organ that acts as a reservoir for urine until it can be expelled from the body. Cystitis is a common infection of the bladder, most commonly affecting women. In the event of bladder discomfort or painful (frequent) urination, recommendations include a urinalysis to rule out infection, increased intake of fluids, acidifying the urine (as with cranberry juice) and supporting the immune system.

Enzyme Supplementation Suggestions:

* **High Potency Digestive Enzyme Formula** with every meal
* **pH Balancing Formula** three times daily between meals
* **High Protease Enzyme Formula** three times daily between meals

BLOOD CLEANSE

The blood's main function is to act as a transport system, but it also has a major role in the defense against infection. The white blood cells are our defense against infection by viruses, bacteria, fungi and parasites as well as any inflammation. Hemoglobin, a protein present in red cells, helps carry oxygen from the lungs to the tissues, where it is exchanged for carbon dioxide. Almost half the volume of blood consists of red and white blood cells, while the remainder is fluid called plasma, which contains dissolved proteins, sugars, fats and minerals. The protease in High Protease Enzyme Formula fortifies and cleanses the blood by helping to build the immune system and break down the unwanted proteins in the blood.

Enzyme Supplementation Suggestions:

- **High Potency Digestive Enzyme Formula** with every meal
- **High Protease Enzyme Formula** three times daily between meals
- **pH Balancing Formula** three times daily between meals

BONE FRACTURE

A fracture is a break in a bone. Therapeutic recommendations include support of the digestive system to optimize digestion and absorption of essential nutrients. The addition of mineral supplementation to aid in bone and tissue regeneration would also be helpful.

Enzyme Supplementation Suggestions:

- **High Potency Digestive Enzyme Formula** with every meal

- **Anti-inflammatory Enzyme Formula** three times daily (more may be taken as needed)

BRONCHITIS

Both bacteria as well as viruses can cause bronchitis, an inflammation of the bronchi (the airways that connect the trachea to the lungs). Chronic bronchitis typically results from repeated lung irritation and may be caused by allergies. It is an ailment that is common among smokers and in individuals living in areas of heavy atmospheric pollution. Acute bronchitis generally develops following upper respiratory tract infections and can develop into pneumonia. Those who suffer from chronic bronchitis have a higher risk of developing heart disease. Symptoms can include difficulty breathing, fever, sore throat, a mucous build-up, and coughing.

Recommendations include determining the underlying cause of the condition, reducing congestion and inflammation, and supporting the digestive and immune systems.

Enzyme Supplementation Suggestions:

- **High Potency Digestive Enzyme Formula** with every meal
- **Seaprose / Mucolase Enzyme Formula** three times daily between meals
- **High Protease Enzyme Formula** three times daily between meals

BRUISING (SEE SOFT TISSUE INJURY)

BURSITIS

Bursitis is the inflammation and tenderness of the sac-like membranes (bursa), which lubricate the joints, tendons, muscles and bones. This condition may result from injury, strain, infection, calcium deposits, or arthritis. The lower knee, hip, shoulder and elbow are the most common locations of inflammation. Often acute pain limits the range of motion as well.

Recommendations include rest, compression of the area and ice to limit swelling, elevation of the injured area to encourage the drainage of fluids, and nutritional support to aid healing.

Enzyme Supplementation Suggestions:

- **High Potency Digestive Enzyme Formula** with every meal
- **Anti-inflammatory Enzyme Formula** three times daily more may be taken as needed
- **High Protease Enzyme Formula** or **Nattokinase Formula** three times daily

CANCER

Over one hundred types of cancer are discernable, all having different causes and rates of aggression. Research indicates that some contributing factors include inflammation, stress, poor diet, hereditary and environmental factors, exposure to certain toxins and chemicals, and a compromised immune system. No one knows for certain why cancer cells and tumors develop in the body.

There are four categories of cancer. Sarcomas affect the connective tissue, muscles and bones; carcinomas affect the skin, glands, organs and mucous membranes; lymphomas affect the lymphatic system, and leukemias affect the blood.

Many different forms of treatment are available, including chemotherapy, drugs, radiation, nutritional support and immune system stimulation and support. Because there are so many different types of cancer all requiring different types of intervention, therapeutic recommendations include working in conjunction with the primary physician and offering nutritional and immune system support.

Depending upon the type of cancer, the condition of the patient and the drug therapies recommended, as well as dietary recommendations should be discussed with the primary physician.

Enzyme Supplementation Suggestions:

- **High Potency Digestive Enzyme Formula** with every meal
- **High Protease Enzyme Formula** or **Nattokinase Formula** at least three times daily
- **pH Balancing Formula** four times per day between meals
- **Probiotic Formula** daily

Optional: **Anti-inflammatory Enzyme Formula** may be added for pain or inflammation.

CANDIDIASIS

This condition involves the overgrowth of the fungus, Candida albicans, which normally exists in an ecological equilibrium throughout the body. Under certain conditions this fungus proliferates, weakening the immune system, causing infections and a wide host of symptoms including but not limited to vaginitis, depression, congestion, GI tract disturbances, acne, muscle and joint pain, heightened environmental sensitivities, hypothyroidism and adrenal insufficiencies. The use of antibiotics, nutrient and digestive deficiencies, altered bowel flora and impaired liver function can all contribute to Candida overgrowth.

Recommendations include identifying predisposing factors, following a Candida control diet, promoting an effective detoxification and elimination program, and supporting the digestive and immune systems.

Enzyme Supplementation Suggestions:

- **High Potency Digestive Enzyme Formula** or **High Amylase Digestion Formula** with every meal
- **High Cellulase / Candida Formula** three times daily between meals
- **High Protease Enzyme Formula** added in acute or chronic conditions
- **Probiotic Formula** daily

Optional: **pH Balancing Formula** two times daily

CANKER SORES

Canker sores (aphthous ulcers) are painful, persistent and annoying sores on or under the tongue, soft palate, on the inside of the cheeks or lips, or at the base of the gums. The sores can range from the size of a pinhead to as large as a quarter. Canker sores are different from fever blisters (cold sores) from the herpes virus, which are contagious and are usually found on the outside of the lips or on the corners of the mouth.

Recurrent canker sores appear to be related to stress, nutrient deficiencies, hormonal imbalances, poor dental hygiene, Crohn's Disease or other GI tract ailments, and food sensitivities (particularly gluten, milk and chocolate).

Recommendations include reducing stress, eliminating food allergens and correcting nutrient deficiencies to balance the body's minerals and pH.

Enzyme Supplementation Suggestions:

High Potency Digestive Enzyme Formula with
   every meal
High Protease Enzyme Formula added before,
   during and after break out

Optional: **pH Balancing Formula** two times per day between meals

CELIAC DISEASE

This rare disorder is characterized by diarrhea and an

abnormal small intestine structure caused by intolerance to gluten. The intestinal lining becomes damaged and the ability to absorb vitamins and minerals is impaired. Malabsorption generally becomes a serious problem.

Symptoms include diarrhea, nausea, abdominal swelling, foul smelling greasy stools, weight loss, anemia, joint/bone pain and skin rashes. The diagnosis of Celiac Disease is confirmed by a biopsy of the small intestine.

Recommendations include elimination of foods containing gluten or milk products and support of the body's digestive and immune systems.

Enzyme Supplementation Suggestions:

- **High Potency Digestive Enzyme Formula** or **High Lactase Formula** with every meal
- **High Protease Enzyme Formula** three times daily between meals
- **High Cellulase / Candida Formula** three times daily between meals
- **Soothing Digestive Formula** for nausea or intestinal inflammation
- **Probiotic Formula** daily

Optional: **pH Balancing Formula** three times daily

CHOLESTEROL/TRIGLYCERIDES (ELEVATED)

Diet, heredity, and metabolic diseases influence the level of cholesterol in the blood. Cholesterol is an important constituent of body cells. It is also needed in the formation of hormones and in the transport of fats (triglycerides) in the bloodstream to tissues throughout the body. High blood cholesterol can cause the accumulation of fatty tissue on the inner lining of arteries, which often results in heart disease or strokes.

Recommendations include a well-balanced, low fat (not non-fat), no sugar diet, daily exercise, supplemental nutrients to elevate HDL levels, and a monitoring of the cholesterol/triglyceride levels.

Enzyme Supplementation Suggestions:

- **High Potency Digestive Enzyme Formula** or a **High Amylase Digestion Formula** with every meal (for compete break down of sugars)
- **High Lipase Enzyme Formula** three times daily between meals
- **Nattokinase Formula** three times daily (for heart health)

CHRONIC FATIGUE SYNDROME
(See also FIBROMYALGIA)

The symptoms of CFS and FMS are many, and are so very similar to many other disorders, they become very difficult to diagnose. Although CFS and FMS are not diseases, they are syndromes with specific sets of signs and symptoms that occur together. When patterns showing these symptoms are repeated continuously within a six-month period, they are diagnosed as actually being the patient's condition.

The symptoms of Chronic Fatigue Syndrome and Fibromyalgia may include sore throat, low-grade fever, recurrent fatigue, headaches, lymph node swelling, intestinal discomfort, digestive difficulties, emotional stress, depression, and muscle and joint pain. Many are also severely stressed and depressed. FMS is very similar to Chronic Fatigue Syndrome. The only difference is that CFS has the diagnostic requirement of fatigue and FMS has the requirement for musculoskeletal pain.

Sometimes the person doesn't feel like they are physically ill enough to seek professional help. Those who suffer with Chronic Fatigue and Fibromyalgia (CFS and FMS) do not look sick, so they find themselves constantly on the defensive with their family and friends.

Research shows these people experience the lack of the polysaccharolytic enzymes early in life. Polysaccharolytic enzymes are the catalysts that break down carbohydrates. Clinically, it is shown that people with FMS or CFS have digestive problems with carbohydrates (starches). Many also have a lipase deficiency. Without the proper lipolytic enzymes, one will have fatty acid imbalances, and possibly hormonal imbalances.

Some of the latest information on Chronic Fatigue Syndrome comes from the Temple University School of Medicine in Philadelphia. Dr. Suhadolnik, a professor of biochemistry and a member of the University's Institute of Cancer Research and Molecular Biology explains, "All CFS patients tested have a new enzyme, while none of the healthy controls do." CFS patients typically have an inability to control common viruses and an inability to maintain cellular energy. This newly discovered enzyme does not function as well as the normal enzyme in healthy people. He feels this explains why CFS patients have a hard time maintaining the energy for cellular growth.

Besides all of the above information, the one thing that weaves throughout all of the research is a common belief that these people have a genetic predisposition for these syndromes. In 1999, research showed that over five million people in the United States have been diagnosed with Chronic Fatigue Syndrome. It stands to reason that there are many more which have not been properly diagnosed.

Successfully treating CFS requires a comprehensive approach to identifying the underlying causative factors. Recommendations include lifestyle modification to reduce/ manage stress, identification of food allergies, counseling, daily exercise, detoxification and optimum support of the digestive and immune systems.

Enzyme Supplementation Suggestions:

- **High Potency Digestive Enzyme Formula** add Soothing Digestive Formula if necessary
- **Anti-inflammatory Enzyme Formula** three times daily for inflammation (more may be added)
- **High Protease Enzyme Formula** three times daily
- **pH Balancing Formula** three times daily
- **High Amylase / Allergy Formula** may be added any time to increase energy level

CIRCULATION (POOR)

The continuous flow of blood throughout the body provides all of the body tissues with a regular supply of oxygen and nutrients, and carries away waste products. Therapeutic considerations of poor or restricted blood circulation should include daily exercise and support of the digestive system to optimize the digestion and absorption of nutrients. Optimum blood flow and maintaining the integrity of the cardiovascular system are essential.

Enzyme Supplementation Suggestions:

- **High Potency Digestive Enzyme Formula** with every meal
- **Nattokinase Formula** two times daily between meals

Optional: **High Protease Enzyme Formula** may replace **Nattokinase Formula**

COLDS

Almost 200 viruses, all broadly similar, are known to cause colds. Symptoms of colds include sneezing, coughing, headache, fever, head congestion, restlessness, and generalized aches and pains.

Recommendations include rest and support of the digestive and immune systems.

Enzyme Supplementation Suggestions:

- **High Potency Digestive Enzyme Formula** with every meal
- **High Protease Enzyme Formula** three to six times a day
- **Seaprose / Mucolase Enzyme Formula** three times daily between meals (For mucus, if necessary)

COLIC (INFANTILE)

Colic is thought to be due to spasms in the intestines of a young child. Colic isn't a disease, but a pattern of excessive crying with no apparent cause. Colic may be triggered by a milk/dairy allergy or intolerance, yet the research is not conclusive.

Infantile colic is common, occurring in approximately 1 in 10 babies, starting between the third and sixth week of life and continuing until the child is three or four months old. The baby usually cries, turns red in the face, may pass gas, clench their fists, and draw up their legs.

If the baby runs a fever or becomes ill with bouts of colic, a doctor should be consulted. Over stimulation, rapid changes, coupled with the anxiety of the parents will often make the child even more irritable.

Recommendations include creating a calm, quiet environment during and following mealtimes. Nursing mothers should not eat foods that may be contributing to GI irritability in their infants. For example, dairy products, onions, wheat, and broccoli may need to be eliminated. Administering digestive enzymes to the infant and mother before feeding times may also be highly beneficial.

Enzyme Supplementation Suggestions:

½ cap of Digestive Enzyme Formula mixed in room temperature water through a syringe into the child's mouth with every meal or bottle. If results are not satisfying, increase dosage or repeat as often as needed.

<u>Nursing Mothers:</u> **High Potency Digestive Enzyme**

Formula with meals and before feedings.

Rashes: If a rash develops, the mother should add a **High Cellulase / Candida Formula** three times per day to her diet.

PLEASE NOTE: Do not mix enzymes with milk or with formula, since the enzymes will begin to hydrolyze it. Administer enzymes in water to the infant before the child ingests any food.

COLITIS

Colitis is an inflammation of the mucous membranes of the colon (large intestine). Poor eating habits, food allergies, stress, or a viral or bacteria infection may cause colitis. The symptoms include abdominal cramps and diarrhea, (usually with blood and mucus) and may also be accompanied by fever.

A diet rich in fresh vegetables, fruit and good sources of protein is suggested. Too many starches, fats, red meat or dairy products may irritate the colon. Regular physical activity is very important to keep the gastrointestinal tract functioning normally.

Recommendations also include an elimination diet to rule out food allergens, increased hydration, relaxation to reduce stress, and supplementation with digestive enzymes, probiotics and nutrients to support the immune system.

Enzyme Supplementation Suggestions:

- **High Potency Digestive Enzyme Formula** with meals
- **Anti-inflammatory Enzyme Formula** three times daily (or any time inflammation occurs)
- **High Protease Enzyme Formula** three times daily
- **pH Balancing Formula** three times daily
- **Probiotic Formula** at least two times daily

CONSTIPATION

Constipation is the infrequent or difficult passing of hard, dry feces. It is important that the bowels move daily, ideally after each meal. Harmful toxins can form as a result of waste products that remain in the colon for longer than four hours. As a result of constipation, other difficulties may arise including indigestion, hemorrhoids, piles, obesity, diverticulitis, appendicitis, hernias, and cancer.

Recommendations include increasing hydration, increasing fiber in the diet, regular exercise, and supplementation with digestive enzymes and probiotics to enhance digestion and absorption of nutrients to maintain healthy intestinal flora.

Enzyme Supplementation Suggestions:

- **High Potency Digestive Enzyme Formula** with every meal
- **High Cellulase / Candida Formula** three times daily between meals

Optional: **pH Balancing Formula** two times daily

CRAMPS (Muscle)

Muscle cramps are generally caused by calcium and magnesium imbalance and/or a vitamin E deficiency. They can also be caused by poor circulation. Most muscle cramps occur at night, affecting primarily the calf muscles in the legs.

Recommendations include maintaining a well-balanced diet with vitamin and mineral supplementation, enzymes to enhance absorption and digestion of nutrients, massage and heat to relieve discomfort, and in cases of frequent cramping, a physical evaluation to rule out impaired circulation.

Enzyme Supplementation Suggestions:

- **High Potency Digestive Enzyme Formula** with every meal
- **pH Balancing Formula** two times daily
- **Anti-inflammatory Enzyme Formula** any time for pain relief

CROHN'S DISEASE

Crohn's Disease is a chronic inflammatory disease that can affect any part of the GI tract, from the mouth to the anus. The intestinal wall can become extremely thick due to chronic inflammation. As the inflamed areas heal, scar tissue can narrow the intestinal passageway. This condition may cause pain, fever, diarrhea (sometimes bloody), weight loss, abdominal pain, malabsorption, anemia, and fatigue. The cause of Crohn's Disease is still unknown; however, the risk increases with a history of food allergies.

Recommendations include increased water intake, relaxation to reduce stress and assessment of possible food allergies. Stress may make the problem worse and it may help to drink a calming herbal tea in the evening before bed.

Enzyme Supplementation Suggestions:

- **High Potency Digestive Enzyme Formula** before and after meal
- **High Cellulase / Candida Formula** three times daily between meals
- **Anti-inflammatory Enzyme Formula** three times daily between meals
- **Probiotic Formula** at least two times daily
- **Soothing Digestive Formula** as needed

CYSTIC FIBROSIS (CF)

A congenital metabolic disorder in which secretions of exocrine glands are abnormal; excessively viscid mucus causes obstruction of passageways (including pancreatic and bile ducts, intestines, and bronchi), and the sodium and chloride content of sweat are increased throughout the patient's life; symptoms usually appear in childhood and include meconium ileus, poor growth despite good appetite, malabsorption and foul bulky stools, chronic bronchitis with cough, recurrent pneumonia, bronchiectasis, emphysema, clubbing of the fingers, and salt depletion in hot weather.

CF is a genetic disorder that affects the respiratory, digestive and reproductive systems. There are approximately 30,000 people in the United States with CF, 3,000 in Canada and 30,000 in other areas of the world. In the U.S., there are approximately 2,500 new cases diagnosed each year. Currently, there is no cure for CF, an article in the October

11, 2000; issue of The Journal of the American Medical Association (Wang et al) discusses a possible connection between the gene responsible for cystic fibrosis and chronic rhinosinusitis. Rhinosinusitis refers to inflammation of the mucous membranes that line the nose and facial sinuses. The facial sinuses are the cavities that surround the nose and are part of the upper respiratory tract. CF is an inherited (genetic) condition affecting the glands that produce mucus, tears, sweat, saliva and digestive juices.

Enzyme Supplementation Suggestions:

- **High Potency Digestive Enzyme Formula** with every meal
- **High Protease Enzyme Formula** three times daily between meals
- **Seaprose / Mucolase Enzyme Formula** three times daily

DANDRUFF

This chronic scalp disorder is the result of dysfunctional sebaceous glands in the scalp, forming scales that may burn and itch. Dandruff can be caused by many factors including poor diet and a mineral, nutrient, fatty acid deficiency. It can also be a symptom of Candidiasis. It is important to recognize the difference between true dandruff and just a build-up of the use of hair products, however. Although dandruff is not contagious and is rarely a serious ailment, it can be embarrassing and surprisingly persistent.

Enzyme Supplementation Suggestions:

- **High Potency Digestive Enzyme Formula** with every meal

High Cellulase / Candida Formula three times daily between meals

High Lipase Enzyme Formula three times daily between meals with Essential Fats

High Protease Enzyme Formula three times daily between meals

DECUBITUS ULCERS [See BED SORES]

DEPRESSION

Clinical depression has symptoms that may include insomnia or hypersomnia, physical hyperactivity or inactivity, feelings of worthlessness, loss of interest or pleasure in usual activities, lack of energy, irritability, either poor appetite and weight loss, or increased appetite and weight gain, a diminished ability to think or concentrate and/or recurrent thoughts of death or suicide.

Depression may be caused by stress, nutritional deficiencies, allergies, thyroid disorders, sugar, battling a serious physical disorder, and/or chronic fatigue syndrome. Some people also become more depressed in the winter months when there is less exposure to sunlight. It is estimated that seven million Americans suffer from depression each year and over eight million Americans take antidepressant drugs.

Recommendations include determination of the factor(s) that are contributing to the depression. Eliminating allergens, adding regular physical exercise and seeking psychological support and/or counseling should also be considered.

Enzyme Supplementation Suggestions:

- **High Potency Digestive Enzyme Formula** or a High Amylase Digestion Formula with every meal
- **High Amylase / Allergy Formula** any time to increase energy levels
- **High Protease Enzyme Formula** two times daily between meals

<u>Optional:</u> **pH Balancing Formula** two times daily

DERMATITIS

Also known as eczema, dermatitis is a red, itchy, weepy reaction where the skin has come into contact with a substance that the immune system recognizes as foreign. This condition is an allergy that creates flaking, color change, thickening, scaling, and itching of the skin. Causes may include cosmetics, perfumes, latex or rubber sensitivity, metal alloys (including silver, nickel and gold), and poisonous plants such as poison ivy or poison oak. Dental cavitations have also been known to cause outbreaks of dermatitis on the face and neck.

Recommendations include removal of the irritant, support of the digestive and immune systems and detoxification.

Enzyme Supplementation Suggestions:

- **High Potency Digestive Enzyme Formula** with every meal

- **High Protease Enzyme Formula** three times daily between meals
- **Probiotic Formula** daily

Optional: **High Amylase / Allergy Formula** three times daily between meals

DIABETES

The chronic disorder of diabetes is characterized by elevated fasting blood glucose levels of protein, carbohydrate and fat metabolism. Diabetics are more susceptible to developing heart and kidney disease, strokes and a loss of nerve function.

There are two types of diabetes: diabetes mellitus and diabetes insipidus. Diabetes insipidus is a very rare metabolic condition caused by a deficiency of the pituitary hormone. The symptoms are intense thirst and the excretion of large amounts of urine. Diabetes mellitus occurs as a result of insufficient production of insulin by the pancreas. This creates high levels of glucose in the blood and low levels of glucose absorbed by the tissues.

Additionally, there are two types of diabetes mellitus, type I and type II. Type I is insulin dependent diabetes mellitus, which occurs primarily in children and young adults. Type II is referred to as "maturity onset" or "adult onset" diabetes, which generally occurs later in life in those with a family history of diabetes. Ninety percent of diabetics are type II. Contributing factors to the development of this disorder include obesity, genetic predisposition, environmental conditions, nutritional deficiencies, viral infections and certain chemical exposures that create autoimmunity.

Recommendations include management by a physician for glucose. Vitamins, minerals, and supplementation with glandulars may help prevent complications.

PLEASE NOTE: Under no circumstances should an individual with diabetes be suddenly taken off of insulin or diabetic drugs. Any nutritional supplementation should be administered only under the close guidance and recommendation of the primary physician.

Enzyme Supplementation Suggestions:

- **Digestive Enzyme Blend** with every meal
- **High Lipase Enzyme Formula** with every meal
- **High Protease Enzyme Formul**a two times per day between meals
- **pH Balancing Formul**a two times per day between meals
- **High Potency Glucoreductase Formula** after each meal or sweet snack

DIARRHEA

Diarrhea is characterized by increased fluidity, frequency (more than four times a day), or volume of bowel movements, as compared to what should be a person's normal pattern of excretion. Some individuals may run a fever with diarrhea as well. Acute diarrhea affects almost everyone from time to time. Chronic diarrhea or presence of blood in the stool may be due to a serious intestinal disorder and a physician should be consulted.

The most common causes include food poisoning,

viruses, bacteria, rancid foods, stress, incomplete digestion of food, excessive use of laxatives, antacids, caffeine, parasites, or chemicals that the body cannot tolerate.

Recommendations include increased water intake and replacements of electrolytes. Avoid dairy products and solid foods, supplementation with probiotics to re-colonize the friendly intestinal flora and appropriate treatment and supplementation based on identification of the cause of chronic diarrhea.

Enzyme Supplementation Suggestions:

- **High Potency Digestive Enzyme Formula** with food
- **Soothing Digestive Formula** after each episode
- **Probiotic Formula** three times a day

DIVERTICULOSIS / DIVERTICULITIS

Diverticulitis is a condition in which diverticuli in the colon rupture, resulting in infection in the tissues that surround the colon. A patient suffering from diverticulitis may exhibit abdominal pain, tenderness and fever.

Diverticulosis refers to the presence of diveticulae, small sacs branching out from a hollow organ or structure, such as the intestine, caused by the protrusion of the inner lining of the colon into areas of weakness in the colon wall. This condition is characterized by inflammation, impaction, or perforation of the diverticula.

Typically, individuals with diverticulosis may not have physical symptoms, yet symptoms can include tenderness on the left side of the abdomen that is relieved by a bowel movement or passage of gas, cramping, nausea or constipation or diarrhea. Contributing factors include stress, obesity, poor diet and eating habits, family history, gallbladder disease and coronary artery disease.

Recommendations include a well-balanced, high-fiber diet, stress management, increased water intake, regular exercise, supplementation with probiotics to colonize the GI tract. and nutritional support to optimize digestion and absorption of nutrients.

Enzyme Supplementation Suggestions:

- **High Potency Digestive Enzyme Formula** or **Digestive Enzyme Blend** with every meal
- **Probiotic Formula** daily
- **Anti-inflammatory Enzyme Formula** for inflammation as needed
- **pH Balancing Formula** three times daily between meals

DYSPEPSIA [See INDIGESTION]

EAR INFECTION

Infection of the middle or outer ear can cause pressure to build up in tiny spaces that can place pressure on delicate nerve endings causing pain. The most common type of ear infection is swimmer's ear (Otitis externa). Symptoms include discharge from the ear, slight fever, a pain that intensifies when the ear is pulled or touched and possibly a temporary loss of hearing.

Otitis media or middle ear infections are most common in infants and young children. Symptoms include a fever as high or higher than 103 degrees, earache, pulling at the ears and a feeling of fullness or pressure in the ear. This infection can be triggered by decompression in air travel, increased exposure to cold climates and high altitudes, exposure to parental smoking, wood-burning stoves, and a lowered resistance due to an allergy or upper respiratory infection.

Recommendations include well-balanced diet, ruling out food allergies that can compromise the immune system and nutritional supplementation to improve overall digestion, absorption of nutrients and a strengthening of the immune system.

Enzyme Supplementation Suggestions:

- **High Potency Digestive Enzyme Formula** with every meal
- **High Protease Enzyme Formula** three times daily on an empty stomach

PLEASE NOTE: Use reduced dosages for younger children.

ECZEMA

Eczema (atopic dermatitis) is characterized by a variety of anatomical and physiological abnormalities of the skin, which can include skin that is very dry, itchy, red and scaly, skin that is inflamed and accompanied by blisters and scaling, and/or a thickening of the skin in response to chronic scratching and rubbing. Research indicates that most cases of eczema are due to food allergies.

Recommendations include improving the overall digestion and absorption of nutrients with enzymes and nutritional supplements, supporting the immune system, and when necessary, a detoxification of yeast overgrowth.

Enzyme Supplementation Suggestions:

- **High Potency Digestive Enzyme Formula** with every meal
- **High Cellulase / Candida Formula** and **High Protease Enzyme Formula** three times daily between meals
- **High Lipase Enzyme Formula** three times daily between meals with essential fats
- **Probiotic Formula** two times daily on an empty stomach

EDEMA (Water Retention)

Water accounts for more than half of our body weight and is in constant exchange between the blood and tissues. Water is forced out of the capillaries and into the tissues by the pressure of blood being pumped throughout the body. Through a reverse process that depends on the water-drawing power of proteins in the blood, the capillaries from the tissues reabsorb water.

Maintaining this balance is the action of the kidneys, which pass excess salt from the blood into the urine to be excreted from the body. Until the excess fluid in the body increases by more than five percent, you may only experience a gain in weight. However, once that five percent is reached, edema becomes evident as swelling in various parts of the body. Specific nutritional supplementation is dependent upon ascertaining the underlying imbalance that is creating this condition.

Enzyme Supplementation Suggestions:

- **High Potency Digestive Enzyme Formula** with every meal
- **pH Balancing Formula** three times daily between meals
- **High Protease Enzyme Formula** three times daily between meals

ENDOCRINE GLANDS (Support)

All of the glands in our body require nutritional support and replenishment. Their ongoing work includes production of hormones, circulation, regulation of metabolic activities and nutrient levels, pH balance, stress management, secretion of enzymes to aid in digestion, and development of sexual characteristics. When our body becomes compromised in any way, it can become depleted of vital nutrients that sustain the health of our glands. When one gland is not functioning optimally, the health and activities of all of the other glands can also be affected.

Recommendations include detoxification of any underlying toxins, nutritional therapies, and enzymes to support the digestive and immune systems, lowering of stress levels, and proper rest/sleep.

Enzyme Supplementation Suggestions:

- **High Potency Digestive Enzyme Formula** with every meal
- **pH Balancing Formula** three times daily between meals
- **High Protease Enzyme Formula** three times daily between meals

ENERGY / ENDURANCE

Our diet comprises three main sources of energy: carbohydrates, proteins and fats. All activities of the body require energy, and all needs are met by the consumption of foods containing energy in a chemical form. Carbohydrates are most readily available for the energy to activate muscles. Proteins work to build and restore body tissue. Fats are the most concentrated form of energy.

In a healthy body, it is important to maintain a balance of these energy sources. A lack of energy and endurance can be caused by numerous factors which may include poor diet, stress, malabsorption of nutrients, a lack of enzymes, free radical damage, and hormonal imbalances.

Enzyme Supplementation Suggestions:

- **High Potency Digestive Enzyme Formula** with every meal

- **High Amylase / Allergy Formula** any time a boost in energy is needed
- **Anti-inflammatory Enzyme Formula** before and after physical activity
- **High Protease Enzyme Formula** three times daily between meals

ENVIRONMENTAL TOXICITY

Today's world is filled with a countless array of chemicals, toxins, radiation and heavy metals that have contaminated our water, air and food. When the body is constantly exposed to and threatened by these environmental toxins, the immune system can become depleted. A compromised immune system, in turn, can contribute to the development of imbalances and disease in the body.

Recommendations include: hair, urine and/or blood assays to determine the specific underlying toxin; the supplementation of enzymes, nutrients and antioxidants to support the digestive and immune systems; the establishment of a health diet; stress management; enough rest/sleep; regular physical exercise; the detoxification of the specific toxin(s) through sauna therapy and drainage remedies to expedite the release of those toxins from the body.

Enzyme Supplementation Suggestions:

- **High Potency Digestive Enzyme Formula** with every meal
- **High Protease Enzyme Formula** three times daily between meals
- **pH Balancing Formula** three times daily between meals

EPSTEIN BARR VIRUS [See also CHRONIC FATIGUE SYNDROME and FIBROMYALGIA]

The Epstein Barr Virus is a herpes virus that has the ability to remain in the body long after the initial infection. In circumstances when the immune system is compromised, the EBV can become reactivated repeatedly, compromising and disrupting immunity leading to other disorders such as Chronic Fatigue Syndrome.

Recommendations include optimizing and supporting the immune system, viral detoxification and establishing a healthy, well-balanced diet. Rule out food allergies that tax the immune system. Ensure enough rest and sleep.

Enzyme Supplementation Suggestions:

- **High Potency Digestive Enzyme Formula** with every meal
- **High Protease Enzyme Formula** three times daily (maintenance - full symptom load)
- **pH Balancing Formula** three times daily between meals

Optional: **Anti-inflammatory Enzyme Formula** may be added for inflammation or **High Amylase / Allergy Formula** may be added for increased energy

ERECTILE DYSFUNCTION [See IMPOTENCE]

EXERCISE

Exercise is defined as a performance of physical exertion for improvement of health. It should be a vital and significant part of everyone's life. Talk to your physician about how much exercise is right for you. A reasonable goal for many people is to work up to exercising four to six times a week for 30 to 60 minutes at a time. Remember, though, that exercise has so many benefits that any amount is better than none. Start out slowly. If you've been inactive for years, you can't run the Boston Marathon after two weeks of training. Begin with a 10-minute period of light exercise, or a brisk walk every day. Then gradually increase how hard you exercise and for how long.

The benefits of regular exercise include reducing your risk of heart disease, high blood pressure, osteoporosis, diabetes and obesity; keeping joints, tendons and ligaments flexible; contributes to mental well-being; helps relieve stress and anxiety; helps maintain a normal weight by increasing metabolism.

Enzyme Supplementation Suggestions:

- **High Potency Digestive Enzyme Formula** with every meal
- **Anti-inflammatory Enzyme Formula** before and after physical activity
- **pH Balancing Formula** two times per day morning and night
- **High Protease Enzyme Formula** two times per day morning and night

EYE CONDITIONS

Like all other parts of the body, the eyes require special nutrients in order to remain healthy. Additionally, the health and appearance of the eyes can also serve as a strong indicator of imbalances in other parts of the body. For example:

Thyroid imbalances may be indicated by bulging or protruding eyes.

Allergies may be indicated by swelling, redness, and irritation of the eyes, and/or dark circles under the eyes.

Colds may be indicated by watery eyes.

Liver or gallbladder disease may be indicated by yellowing eyes.

Diabetes/hypertension may be indicated by periodic blurred vision.

Therapeutic considerations of eye conditions include avoiding eyestrain, eating a well-balanced diet, supplementation of the diet with antioxidants to protect against free radical damage, and the addition of enzymes or supplemental nutrients to help maintain health, and a viable digestive and immune system.

It is always wise to consult with a physician in cases of infections or serious eye imbalances such as pinkeye, shingles, or ulcerations near the eye, glaucoma, macular degeneration and detached retinas.

Enzyme Supplementation Suggestions:

- **High Potency Digestive Enzyme Formula** with every meal

- **High Protease Enzyme Formula** and **Natto-kinase Formula** three times daily to increase circulation

FATIGUE

Fatigue is a common complaint that can be associated with overwork or lack of sleep. Persistent fatigue may either be a result of, or result in depression, lowered immune system response, headaches, inflamed throat and lungs, fevers, and swollen lymph glands.

Recommendations include eating a well-balanced diet, time for an adequate amount of rest and sleep, stress management, ascertaining the underlying cause of the fatigue, and support of the digestive and immune systems with nutritional supplementation and enzymes. For the overall health of the body, these suggestions cannot be emphasized enough.

Enzyme Supplementation Suggestions:

- **High Potency Digestive Enzyme Formula** add **Soothing Digestive Formula** if necessary
- **Anti-inflammatory Enzyme Formula** three times daily for inflammation (more may be added)
- **High Protease Enzyme Formula** three times daily
- **pH Balancing Formula** three times daily
- **High Amylase / Allergy Formula** may be added any time to increase energy level

FEVER

A fever is a rise in body temperature to greater than 100 degrees. The body maintains stability between 97 degrees and 100 degrees by balancing the heat produced by the body's metabolism with the heat lost to the environment. The "thermostat" is located in the hypothalamus within the brain. A fever occurs when the body's thermostat resets at a higher temperature, primarily in response to an infection.

A fever may be accompanied by other symptoms such as shivering, headache, sweating, thirst, a flushed face, hot skin and faster than normal breathing. Fever symptoms indicate the presence of an imbalance or disease process. An elevated temperature can be helpful to the body since its serves to destroy foreign microbes, assisting the body to eliminate toxins. High fevers may cause delirium, confusion, dehydration, seizures or even a coma. A physician should immediately be consulted with a persistent high fever.

Recommendations include increasing hydration, rest/sleep, avoiding heavy or solid foods until the condition has improved, and supporting the immune system.

Enzyme Supplementation Suggestions:

- **High Potency Digestive Enzyme Formula** with every meal
- **High Protease Enzyme Formula** or **Anti-inflammatory Enzyme Formula** or combination of both, three times daily
- **pH Balancing Formula** three times daily

FIBROCYSTIC BREAST DISEASE

Fibrocystic Breast Disease (FBD) is generally a component of premenstrual syndrome. This condition is characterized by multiple cysts in the breast tissue that become filled with fluid, giving the breast a nodular consistency. The fibrous tissue surrounding the cysts thicken like a scar. These physiological changes often create pressure, tenderness and pain that are usually cyclic in nature.

It is estimated that nearly 50 percent of adult females have FBD. Research indicates that numerous factors can contribute to this condition, including stress, chronic constipation, hormonal imbalances, the ingestion of methylxanthines (which include caffeine, coffee, tea, chocolate, cola and caffeinated medications), hypothyroidism, stress, and faulty liver function.

PLEASE NOTE: Consult with a physician immediately if you should notice a breast lump of any kind.

Enzyme Supplementation Suggestions:

- **High Potency Digestive Enzyme Formula** with **every meal**
- **High Protease Enzyme Formula** and **Nattokinase Formula** three times daily
- **High Lipase Enzyme Formula** three times daily between meals with essential fat
- **pH Balancing Formula** three times daily between meals

FIBROMYALGIA
[See also CHRONIC FATIGUE SYNDROME]

Fibromyalgia disorder is a common cause of chronic musculoskeletal pain and fatigue affecting about 4% of the population. Research indicates that people with FMS have defects in the neuroregulatory system. This abnormal production of the neurotransmitters such as seratonin, melatonin, dopamine and other chemicals that help control pain, mood, sleep and the immune system are the reason for so many of their symptoms. One of the key findings in FMS is an altered sleep pattern with reduced REM sleep and increased non-REM sleep.

Symptoms of this disorder include six or more typical, reproducible tender points in the body, joint swelling, generalized stiffness or aches of at least three anatomical sites for at least three months, sleep disturbances, generalized fatigue, numbness or tingling, IBS, chronic headache and/or neurological and psychological complaints. While the severity of symptoms fluctuate from person to person, FMS may resemble a post-viral state. FMS is very similar to Chronic Fatigue Syndrome. The only difference is that CFS has the diagnostic requirement of fatigue and FMS has the requirement for musculoskeletal pain.

Enzyme Supplementation Suggestions:

- **High Potency Digestive Enzyme Formula** with **every meal**
- **Anti-inflammatory Enzyme Formula** three times daily or anytime with pain
- **High Protease Enzyme Formula** two times daily morning and night

- **pH Balancing Formula** three times daily morning and night

FLATULENCE

Flatulence is the expulsion of intestinal gases formed by fermentation, formed in the gastrointestinal tract as a result of the action between bacteria, carbohydrates and proteins in undigested foods. It is released under pressure through the anus, often with a characteristic sound and offensive odor. It is also often accompanied by abdominal discomfort. Large amounts of flatus (gases) that cause abdominal discomfort are usually due to intestinal sensitivities. [For more information see IRRITABLE BOWEL SYNDROME.]

Enzyme Supplementation Suggestions:

- **High Amylase Digestion Formula** with every meal (more may be taken if necessary)
- **pH Balancing Formula** three times daily

FLU [See INFLUENZA]

FUNGUS

Yeast (fungus) or mold can infect numerous areas of the body, including the mouth, vagina, skin, nails and feet. A compromised immune system is a major cause of fungal infections. Individuals who may have a higher risk of developing yeast infections include diabetics, those suffering from chronic diseases, individuals who are obese, those who take antibiotics, and oral contraceptive users.

Enzyme Supplementation Suggestions:

- **High Potency Digestive Enzyme Formula** or **High Amylase Digestion Formula** with every meal
- **High Cellulase / Candida Formula** three times daily between meals
- **High Protease Enzyme Formula** added in acute or chronic conditions
- **Probiotic Formula** daily

Optional: **pH Balancing Formula** two times daily

GALLBLADDER IMBALANCES

The gallbladder is a small muscular sac that lies under the liver and expels bile through the common bile duct into the duodenum. Gallstones or gallbladder inflammation are the most common disorder of the gallbladder and may cause a great deal of discomfort, primarily in the upper right abdominal region, often as a result of eating fatty or fried foods. Other symptoms may include nausea, vomiting and fever. It is important to consult with a physician in cases of cholecystitis (inflammation of the gallbladder), as untreated cases can be life threatening.

Recommendations include enzymes and nutritional support to aid in proper fat metabolism, eating a health well-balanced diet, and a detoxification program for the liver and colon to improve gallbladder function.

PLEASE NOTE: If a smaller dose of enzymes than suggested is consumed, it may result in nausea.

Enzyme Supplementation Suggestions:

ACUTE:
- **High Potency Digestive Enzyme Formula** with every meal
- **High Lipase Enzyme Formula** between meals

CHRONIC:
- **High Potency Digestive Enzyme Formula** and **High Lipase Enzyme Formula** with every meal
- **High Lipase Enzyme Formula** three times daily between meals
- **Soothing Digestive Formula** whenever nauseous

GASTRITIS

Gastritis is a medical term for an inflammation of the mucous membrane that lines the stomach. The illness may be acute, occurring as a sudden attack, or chronic, developing gradually over a longer period of time. The cause of the irritation may be from infection, bile reflux, stress, or excessive consumption of drugs, alcohol, aspirin, or tobacco. It produces many of the same symptoms as a gastric ulcer.

Recommendations include stress reduction, dietary changes, enzymes and nutritional support to soothe the stomach and GI tract and to enhance digestion and absorption, and elimination of the underlying cause of the gastritis.

Enzyme Supplementation Suggestions:

- **High Potency Digestive Enzyme Formula** with every meal
- **Soothing Digestive Formula** after every meal

GASTROESOPHAGEAL REFLUX (GERD)

This is the backflow of the contents of the stomach into the esophagus that is often the result of incompetence of the lower esophageal sphincter. Gastric juices are acidic and therefore produce burning pain in the esophagus. Repeated episodes of reflux often result in this condition.

The function of the lower esophageal sphincter (LES) is to prevent gastric contents from backing up into the esophagus. Normally, the LES creates pressure, closing the lower end of the esophagus, but relaxes after each swallow to allow food into the stomach. Reflux occurs when pressure within the stomach exceeds LES pressure; it does not occur because there is too much acid in the stomach.

The traditional medical treatment is to prescribe anti-acids. Anti-acids not only mask the problem, they also can cause long-term digestive difficulties. Anti-acids are merely Band-Aids and do nothing to correct the problem.

Enzyme Supplementation Suggestions:

- **High Potency Digestive Enzyme Formula** if ulcers are present with meals
- **Soothing Digestive Formula after every meal**
- **pH Balancing Formula** three times daily

GINGIVITIS [See PERIODONTAL DISORDERS]

An early stage of periodontal disease, gingivitis is a condition that involves an inflammation of the gums. Deposits of food particles, mucus, and bacteria often cause an accumulation of plaque. Symptoms of gingivitis include plaque accumulation around the teeth that cause the tissue to become swollen, red, and infected. Bleeding often occurs. A poor diet, poorly fitted fillings that irritate the gums, and frequent breathing through the mouth can also aggravate or cause gingivitis.

Enzyme Supplementation Suggestions:

- **High Potency Digestive Enzyme Formula** with every meal
- **High Amylase Digestion Formula** may be added to any high carbohydrate or sugar foods
- **High Protease Enzyme Formula** three to four times daily between meals

GOUT

Gout is a metabolic disorder and common form of arthritis that is caused by the accumulation of uric acid crystals in the joints. An intensely painful disease, in most cases, gout affects only one joint, most commonly the big toe, although it can affect other joints: ankle, heel, instep, knee, wrist, elbow or spine. Urate crystals become deposited in tendons, joints, subcutaneous tissue, cartilage, kidneys and other tissues, which then cause inflammation and damage.

Painful gout attacks usually occur at night, may involve fever and chills and are generally preceded by alcohol ingestion, dietary excess, stress, trauma or certain drugs. Some sources report that gout can be a secondary result of a systemic yeast/fungal infestation and lead toxicity.

Recommendations include reducing inflammation, elimination of alcohol and high-purine content foods (meat, fish, lentils, peas, mushrooms, cauliflower, alcohol), reduction of refined carbohydrates and saturated fats, and moderation of protein intake. Other suggestions might include increasing intake of cherries and blueberries which lower uric acid levels, ascertaining the underlying cause of the disorder, supporting the digestive and immune systems, increasing hydration and stress management.

Enzyme Supplementation Suggestions:

- **High Potency Digestive Enzyme Formula** with every meal
- **Anti-inflammatory Enzyme Formula** three times daily or anytime needed for pain
- **High Protease Enzyme Formula** three times daily between meals
- **pH Balancing Formula** three times daily between meals

GUM SUPPORT AND MAINTENANCE

Anyone who has had a gum disorder will tell you the importance of taking care of your teeth and gums. Gum disorders can include gingivitis, bleeding gums and pyorrhea. [See PERIODONTAL DISORDERS]

Recommendations include a healthy well-balanced diet, regular brushing and flossing of the teeth, elimination of toxic agents (smoking, alcohol, excess sugar), and nutritional support.

Enzyme Supplementation Suggestions:

- **High Potency Digestive Enzyme Formula** with every meal
- **High Protease Enzyme Formula** three times daily
- **pH Balancing Formula** three times daily

HAIR LOSS

Loss of hair or baldness (alopecia) may be the result of your DNA or can be caused by many factors that may include illness, skin disease, poor diet, poor circulation, diabetes, thyroid disease, surgery, radiation or chemotherapy, pregnancy (hormonal changes), iron deficiency, drugs and stress.

Enzyme Supplementation Suggestions:

- **High Potency Digestive Enzyme Formula** with every meal
- **High Protease Enzyme Formula** three times daily on an empty stomach
- **pH Balancing Formula** three times daily

HALITOSIS

Halitosis is the medical term for bad breath. This condition may be caused by poor digestion, poor oral hygiene, gum disease, tooth decay, constipation, throat or nose infection, liver insufficiencies, inadequate protein digestion, and/or smoking.

Recommendations: A thorough physical exam is recommended for persistent halitosis to determine the underlying cause. It is very often an indication of toxicity. [See also GUM DISORDERS and PERIODONTAL DISORDERS.]

Enzyme Supplementation Suggestions:

- **High Potency Digestive Enzyme Formula** with every meal
- **High Protease Enzyme Formula** three times daily
- **pH Balancing Formula** three times daily

In addition, consider implementing the complete Cleanse and Fortify program.

HANGOVER

This condition is characterized by headache, nausea, vertigo, diarrhea and depression. The severity of a hangover is determined by the amount and type of alcohol that was consumed. Besides taking supplemental enzymes and maintaining proper nutrition, the obvious recommendation is to not over-consume alcohol.

Enzyme Supplementation Suggestions:

- **High Potency Digestive Enzyme Formula** with every meal
- **High Amylase / Allergy Formula** before, during and after heavy drinking
- **High Lipase Enzyme Formula** anytime symptoms of dizziness or vertigo occur
- **Soothing Digestive Formula** anytime nauseous

HAY FEVER

Hay fever, also referred to as allergic rhinitis, is an allergic response of the nasal passages and airways to air or windborne pollens. This disorder shares many common features with asthma. Nearly 75% of the hay fever conditions in the USA is a result of ragweed pollen. Other causative factors may include feathers, animal hair, dust and pollen from flowering plants, trees and grasses, and food allergens or sensitivities. Symptoms of hay fever may include sneezing, itchy eyes, headaches, nervous irritability and a watery discharge from the nose and eyes.

Recommendations include reducing the allergic response by avoiding the causative airborne and/or food allergens, optimizing the immune system, supporting the adrenal glands, improving overall digestion and absorption of nutrients. (For more information see ASTHMA or ALLERGIES.)

Enzyme Supplementation Suggestions:

- **High Potency Digestive Enzyme Formula** with every meal

- **High Amylase / Allergy Formula** three times daily more may be added for acute reactions
- **High Protease Enzyme Formula** three times daily between meals (for the support of the Immune System)

HEADACHE

Headaches are one of the most common types of pain. The discomfort may either be felt throughout the entire head, or may occur in just one area. Many headaches are the body's response to an adverse stimulus (such as hunger or stress) or a change in the weather. Causes of headaches include but are not limited to tension, illness, pollution, caffeine, dehydration, food allergy/reaction, alcohol, sulfites, fever, drugs, constipation, toxins, vitamin deficiencies, poor vertebral alignment, and hypertension.

There are four types of headaches; vascular, myogenic (muscle tension) traction, and inflammatory. A migraine is a severe, incapacitating headache preceded or accompanied by visual and/or stomach disturbances. Tightening in the muscles of the face, neck and scalp causes tension headaches. Cluster headaches can cause severe pain behind one eye.

Persistent headaches may be due to your environment or food intake. In this case, start with the Cleanse and Fortify Program to bring your body back into balance. After a few days, continue the Optimum Health 2 program to insure good health.

Recommendations include determining the underlying cause of the headaches and avoiding exposure to this element, stress management, getting enough rest/sleep, and supporting the digestive and immune systems

PLEASE NOTE: In the event that your headache is accompanied by blurred vision, heart pounding, visual color changes, sensitivity to light, pressure behind the eyes that is relieved by vomiting, or was the result of a blow or an injury to the head, consult with your physician immediately.

Enzyme Supplementation Suggestions:

- **High Potency Digestive Enzyme Formula** with every meal
- **Anti-inflammatory Enzyme Formula** three times daily or anytime needed for pain

HEARTBURN

Heartburn is a painful or burning sensation in the esophagus, just below the breastbone, caused by regurgitation of gastric acid. It can also be experienced as a burning pain in the center of the chest that may travel from the tip of the breastbone to the throat. It can be the result of overeating, allergies, stress, drinking alcohol or coffee, ulcers, gallbladder problems, and/or eating rich, fried, fatty or spicy foods. Sometimes this pain/discomfort is mistaken for heart problems.

Recommendations include determining the underlying cause of the heartburn and avoiding any causative elements, enzymes (avoid enzymes with HCL) to aid in the proper digestion and absorption of foods, stress management and thoroughly chewing all foods before swallowing.

Enzyme Supplementation Suggestions:

- **High Potency Digestive Enzyme Formula** with every meal
- **Soothing Digestive Formula** after every meal
- **pH Balancing Formula** before bed

HEMORRHOIDS

Hemorrhoids are varicosities or swelling and inflammation of the veins in the rectum (internal hemorrhoids) and anus (external hemorrhoids), accompanied by bleeding when irritated. Factors that can cause the development of hemorrhoids include constipation, poor diet, lack of adequate fluids, and lack of exercise, pregnancy, allergies, liver damage, and heavy lifting. Severe hemorrhoids may require surgery.

Recommendations include determining the cause of the condition, increasing fiber and fluids in the diet, establishing bowel regularity, and supplementation with enzymes to aid in the healing process.

Enzyme Supplementation Suggestions:

- **High Potency Digestive Enzyme Formula** with every meal
- **High Cellulase / Candida Formula** three times per day between meals (for constipation)
- **Nattokinase Formula** or **High Protease Enzyme Formula** three times daily (for circulation)

HERPES VIRUS

The herpes virus is a recurrent viral infection of the skin and mucous membranes. Small, often painful watery blisters that occur around the mouth, lips, genitals, conjunctiva and cornea characterize it.

Herpes Type I (Herpes Simplex) involves skin eruptions, cold sores or fever blisters. It can also cause inflammation of the cornea, which if left untreated, can cause encephalitis.

Herpes Type II (Genital Herpes) is the most prevalent herpes infection. Sexually transmitted, it affects the genitals, involving painful blisters that develop in women around the rectum, clitoris, cervix and in the vagina. Men generally experience blisters that appear on the groin, penis and scrotum. These blisters, which later develop into painful ulcers, are often accompanied by painful urination, swelling and a urethral discharge. Both men and women may experience swollen lymph nodes in the region, muscular aches and a low-grade fever.

The herpes virus becomes dormant in the nerve cells in most individuals after the initial infection. Other individuals, however, can experience recurrent outbreaks. Type I has a recurrence rate of 4%, while Type II has a 60% recurrence rate. Recurrent outbreaks may follow stress, sun exposure, minor infections and trauma. The incubation period for this contagious virus is two to twelve days.

Recommendations include reducing dietary toxicity, emotional and environmental stress, and providing optimum support of the immune system.

Enzyme Supplementation Suggestions:

- **High Potency Digestive Enzyme Formula** with every meal
- **High Protease Enzyme Formula** three times daily between meals
- **pH Balancing Formula** two times daily morning and night

HERPES ZOSTER

This is the medical term for "shingles," a relatively common disease, which is an infection of the nerves of the skin. The most serious feature of the virus is the pain following the rash of small, crusting blisters. After the rash heals, the pain may persist for months, or even years, as a consequence of damage to the nerves. It is possible to reduce the severity of the active stage and to minimize nerve damage by the prompt use of antiviral drugs.

It is caused by the varicella zoster virus, which also causes chickenpox. After chickenpox, most of the viral organisms are destroyed, but some survive and lay dormant in certain sensory nerves. A decline in the efficiency of the immune system allows the viruses to re-emerge and cause shingles.

Herpes zoster can occur under stress, or for those whose immune system has been weakened either by diseases, such as lymphoma or Hodgkin's disease, or by treatment with immuno-suppressant or anticancer drugs. The first indication of this virus is excessive sensitivity in the area of skin to be affected, soon followed by pain, and then the rash appears (after about five days) which can possibly leave scars after crusting.

Recommendations include reducing dietary toxicity, emotional and environmental stress, and providing optimum support of the immune system.

Enzyme Supplementation Suggestions:

- **High Potency Digestive Enzyme Formula** with every meal
- **High Protease Enzyme Formula** three times per day, increase to four times per day at outbreak
- **pH Balancing Formula** three times daily
- **Anti-inflammatory Enzyme Formula** may be added during outbreak (for inflammation and pain)

HIATAL HERNIA [See also GERD]

Hiatal hernia is the protrusion (or hernia) of the upper part of the stomach into the thorax through a tear, opening or weakness (hiatus) in the diaphragm. The muscle at the end of the esophagus may be affected and can cause acid reflux (regurgitation of acidic gastric juices into the esophagus). This reflux, in turn, may cause heartburn and belching that may be made worse by either bending or lying down.

Recommendations include avoiding large, heavy meals, the supplementation of digestive enzymes to promote efficient and complete digestion and absorption of foods, and raising the head of the bed during sleep.

Enzyme Supplementation Suggestions:

- **High Potency Digestive Enzyme Formula** with every meal
- **Soothing Digestive Formula** after every meal and as needed for discomfort

HIGH BLOOD PRESSURE [See HYPERTENSION]

HOT FLASHES [See MENOPAUSE]

HORMONAL IMBALANCES [See ENDOCRINE GLANDS]

HYPERACTIVITY

Hyperactivity involves a disorder of certain CNS mechanisms, which can affect both children and adults. The symptoms may include one or more of the following: lack of concentration, impatience, self-destructiveness, quick frustration, temper tantrums, head knocking, sleep disturbances, clumsiness, and failure in the classroom. Contributing factors include but are not limited to: oxygen deprivation at birth, prenatal trauma, heredity, food allergies, and artificial additives in food, smoking during pregnancy, lead poisoning, foods containing salicylates, a low protein diet, a protease deficiency, and environmental pollutants.

Recommendations include diagnostic tests and evaluations to ascertain the underlying cause of the condition, eating a well-balanced diet and avoiding sugars and foods with a high glycemic index, ruling out or treating sensitivities or allergies to foods, preservatives, additives, and/or environmental toxins. Also recommended are detoxification and drainage products to regenerate the liver, lymph system and blood cleansing. Enzymes and nutritional products are suggested to support the digestive and immune systems.

Enzyme Supplementation Suggestions:

High Potency Digestive Enzyme Formula with every meal and snacks
pH Balancing Formula before bed
High Protease Enzyme Formula three to four times per day (for detoxification)

HYPERTENSION

Hypertension is abnormally high blood pressure. Mild hypertension may respond to weight and/or stress reduction. Factors that increase the risk of this disorder include smoking, obesity, excessive use of stimulants and caffeine, drug abuse, the use of contraceptives, and a high sodium/low potassium intake.

High blood pressure is categorized into different levels: A blood pressure level below 120/80 mmHg (described as 120 over 80) is considered normal. Blood pressure between 120/80 mmHg and 139/89 mmHg is considered *prehypertension* and blood pressure over 140/90 mmHg is considered high.

Recommendations include dietary and lifestyle changes, stress reductions, diagnostic tests and evaluations to determine the underlying cause of the hypertension, enzymes to facilitate digestion and absorption of foods and fats, in addition to supplementation with minerals to correct existing mineral deficiencies.

Enzyme Supplementation Suggestions:

- **High Potency Digestive Enzyme Formula** with every meal
- **Nattokinase Formula** three times daily between meals

HYPOGLYCEMIA

Hypoglycemia literally means "low blood sugar." The principle symptoms arise from an inadequate supply of glucose as fuel to the brain, resulting in an impairment of function. This condition refers to a blood glucose level below 50 mg/dl.

There are two categories of hypoglycemia: fasting and reactive. Fasting hypoglycemia is the result of taking drugs used in the treatment of diabetes. Reactive hypoglycemia, the most common form, involves symptoms that develop three to five hours after meals that are relieved by eating.

Symptoms can include one or more of the following: anxiety, irritability, headache, depression, trembling, excessive sweating, palpitations, confusion, double vision, weakness, bizarre behavior, incoherent speech and convulsions.

Recommendations: The primary goal in hypoglycemia management is to re-establish healthy blood sugar control. This can be achieved by the use of dietary therapies in which frequent small meals are encouraged. In addition, all processed, concentrated, simple carbohydrates and alcohol are avoided.

Enzyme Supplementation Suggestions:

- **High Potency Digestive Enzyme Formula** or **High Amylase Digestion Formula** with every meal
- **High Amylase / Allergy Formula** three times daily (or more as needed) between meals to raise glucose levels and break down excess sugars.

HYPOTHYROIDISM

Hypothyroidism is a disease state caused by insuffi-cient production of thyroid hormone by the thyroid gland. An insufficiency in thyroid hormones can have an effect on all body functions. A deficiency generally results in a large array of signs and symptoms. These can include fatigue, weakness, depression, infertility, constipation, low basal body temperature (below 97.6), thin, brittle nails, hair loss, rough dry skin, weight gain, muscle weakness, forgetfulness, difficulty concentrating, joint stiffness, prolonged and heavy menstrual bleeding, cold hands and feet, sluggish lymphatic drainage, impaired kidney function, and increased risks of heart disease, hypertension and atherosclerosis.

Hypothyroidism can be caused by a deficiency of a thyroid hormone, a lack of pituitary gland stimulation, an iodine deficiency, goiters, and Hashimoto's Disease (an autoimmune disease). Most cases of hypothyroidism are not present at birth, but developed as an adult. Advanced hypothyroidism may cause severe complications.

Recommendations include maintaining a well-balanced diet, getting regular exercise and possibly using a natural thyroid glandular to stimulate thyroid gland secretion.

Enzyme Supplementation Suggestions:

High Potency Digestive Enzyme Formula with
every meal
High Lipase Enzyme Formula three times daily
between meals
High Amylase / Allergy Formula may be added to
help increase energy level

- **High Protease Enzyme Formula** three times per day (for detoxification and organ support)

IMPOTENCE

Impotence is often referred to as erectile dysfunction. It is a sexual dysfunction characterized by the inability to develop or maintain an erection of the penis for satisfactory sexual intercourse regardless of the capacity of ejaculation. This condition currently affects 10-20 million American men, and 5% of men over the age of 50.

Causes of impotence can include vascular insufficiency, stress, certain drugs, neurological diseases, endocrine disorders including diabetes and hypothyroidism, certain diseases or trauma to the male sexual organs, excessive use of alcohol and tobacco products, and psychological factors.

Recommendations include a healthy, well-balanced diet high in quality protein, regular exercise, increasing vitamin and mineral intake and avoid smoking and alcohol.

Enzyme Supplementation Suggestions:

- **High Potency Digestive Enzyme Formula** with every meal
- **pH Balancing Formula** three times daily morning and night
- **Nattokinase Formula** three times daily morning and night (for circulation)
- **High Lipase Enzyme Formula** three times daily morning and night (for hormones)

INDIGESTION

A lack or failure of digestive function can result in indigestion. It is a condition that is frequently caused by eating too fast, especially by eating high-fat or spicy foods quickly. Food that has been badly cooked or over processed can also promote indigestion. In addition, overeating, and eating when tense, tired, or emotionally upset can also aggravate indigestion.

Symptoms may include heartburn, bloating, nausea, flatulence, burping, cramps, a disagreeable taste in the mouth, belching, and sometimes vomiting and diarrhea. A deficiency in digestive enzymes can severely compromise the digestive process, and overtime result in digestive failure.

Recommendations include eating a well-balanced healthy diet, encouraging thorough mastication (chewing) of foods, eating in a comfortable, relaxed environment and avoid drinking large quantities of water during meals.

Enzyme Supplementation Suggestions:

- **High Potency Digestive Enzyme Formula** or **Digestive Enzyme Formula** with every meal
- **Soothing Digestive Formula** after meals
- **High Protease Enzyme Formula** two times daily

INFECTION

Disease-causing microorganisms (such as bacteria, virus and fungi) can establish opportunistic colonies in and on the human body. An infection is the detrimental colonization of a host organism by a foreign species, multiplying at the expense of the host. The host's response to infection is inflammation. In other words, in the body's efforts to control and eliminate these microorganisms, inflammation can arise. This process increases the flow of blood to the infected area, bringing white blood cells and other components of the immune system.

Symptoms can include an elevated temperature, redness, pain, swelling and the formation of pus.

Enzyme Supplementation Suggestions:

- **High Potency Digestive Enzyme Formula** with every meal
- **High Protease Enzyme Formula** three times daily
- **Anti-inflammatory Enzyme Formula** anytime for pain or inflammation
- **pH Balancing Formula** three times daily

INFERTILITY

Infertility is the inability to naturally conceive a child or the inability to carry a pregnancy to term. This condition refers to the inability to conceive a child after a year or more of regular, unprotected sexual activity during the time of ovulation. Nearly 5% of American couples suffer from infertility.

Causes of female infertility may include thyroid disorders, adrenal disease, genetic, cervical or vaginal factors, ovarian or uterine factors, additional hormonal imbalances, Chlamydia, pelvic inflammation, and allergic reactions to their partner's sperm.

Causes of male infertility may include endocrine problems, thyroid disorders, deficient sperm production, inflammation, ductal obstruction, ejaculatory dysfunction, and coital disorders. Approximately 90% of cases of low sperm count have been linked to environmental, lifestyle and dietary changes. There is rising evidence that exposure to increased pollution, toxic solvents and heavy metals including mercury, lead and arsenic are also contributing to falling sperm counts.

Recommendations include ascertaining the underlying cause of the infertility, eating a healthy, well-balanced diet, detoxification, stress management, supplementation with antioxidants and zinc and wearing loose-fitting clothing.

Enzyme Supplementation Suggestions:

- **High Potency Digestive Enzyme Formula** with every meal
- **High Protease Enzyme Formula** three times daily
- **High Lipase Enzyme Formula** three times daily
- **pH Balancing Formula** three times daily

INFLAMMATION

Inflammation is the first response of the immune system to infection or irritation. It is characterized by redness, swelling, heat, pain and dysfunction of the organs or tissues involved. Lymphatic congestion or infection can also result in inflammation. When inflammation occurs, a healthy immune system sends out an accumulation of white blood cells to the site to help destroy invading microorganisms.

Enzyme Supplementation Suggestions:

- **High Potency Digestive Enzyme Formula** with every meal
- **Anti-inflammatory Enzyme Formula** three times daily
- **High Protease Enzyme Formula** three times daily
- **pH Balancing Formula** three times daily

INFLUENZA (Flu)

Influenza is an infection of the lungs and airways with one of the influenza viruses. The symptoms of flu include fever, chills, runny nose, sore throat, cough, headache, GI disturbances, lack of appetite, muscle aches, neuralgia, extreme fatigue, and a general feeling of malaise. Influenza is a highly contagious viral infection of the respiratory tract.

Coughing and sneezing can spread this common ailment also known as the "flu." The incubation period is one to three days. Influenza may also increase susceptibility to pneumonia, sinus problems and ear infections.

Recommendations include increasing hydration, rest, and support of the immune and digestive systems.

Enzyme Supplementation Suggestions:

- **High Potency Digestive Enzyme Formula** with every meal
- **Seaprose / Mucolase Enzyme Formula** three times daily
- **High Protease Enzyme Formula** three to four times per day

- **pH Balancing Formula** three times daily

INSECT BITES

All insect bites provoke a reaction in the skin that is primarily an allergic response to substances in the insect's saliva, venom or deposited feces. A honeybee's stinger is barbed and remains in the wound. The bee sting sac is rich in sugars. The wasp sting is primarily the stomach organ and the stinger of the wasp, which is protein. It is vital to administer these enzymes as soon as possible while the wound is still open.

Enzyme Supplementation Suggestions:

Bee Sting

- **High Amylase / Allergy Formula** open capsule and mix with a very small amount of tepid water. Apply to the sting and reapply after five minutes.
- **High Amylase / Allergy Formula** and **High Protease Enzyme Formula** three times between meals

Wasp Sting

- **High Protease Enzyme Formula** open capsule and mix with a very small amount of tepid water. Apply to the sting and reapply after five minutes.
- **High Protease Enzyme Formula** three times daily between meals

INSOMNIA

Insomnia is defined as habitual sleeplessness, repeated night after night. Most individuals suffering from this ailment have difficulty falling asleep or staying asleep. Insomnia is a symptom, although commonly thought of as a sleep disorder.

Daytime fatigue, irritability and difficulty coping with tasks are other frequent symptoms. Half of the cases of insomnia result from excessive worry, stress, an overactive mind or physical pain. It can also result from certain drugs, hypoglycemia, asthma, fear, anxiety, indigestion, poor diet, and caffeine. Also it can be a result of a deficiency of certain nutrients including calcium, potassium, and magnesium.

Enzyme Supplementation Suggestions:

- **High Potency Digestive Enzyme Formula** with every meal
- **pH Balancing Formula** two times daily, morning and night
- **High Protease Enzyme Formula** two times daily, morning and night

IRRITABLE BOWEL SYNDROME (IBS)

Irritable Bowel Syndrome (often called irritable colon or spastic colon) involves a disturbance of involuntary muscle movement in the large intestine. IBS is a functional bowel disorder characterized by abdominal pain and changes in bowel habits not associated with any abnormalities. The

cause of this ailment is not fully understood.
Symptoms include a combination of intermittent abdominal pain and regular bowel habits, i.e., constipation, diarrhea, or both, alternating. Long-term IBS can contribute to enzyme and nutrient deficiencies, which in turn, can compromise the immune system and interfere with the body's energy production.

Recommendations include dietary changes (including removing starches, high fat foods, and fried foods).

Enzyme Supplementation Suggestions:

* **High Potency Digestive Enzyme Formula** with every meal
* **Soothing Digestive Formula** after every meal and any time there is discomfort
* **Probiotic Formula** daily

JOINT PAIN

A joint is the point at which two bones meet. The ends of the bones are covered with cartilage and the joint is lined with a membrane. Joint pain is the presence of painful joints in the absence of frank arthritis. Common joint injuries include sprains, ligament tears, cartilage damage and tears of the joint capsule. Dislocation of a joint is usually the result of injury, but can be congenital.

Recommendation for joint pain includes supporting the digestive and immune systems, and controlling any inflammation.

Enzyme Supplementation Suggestions:

- **High Potency Digestive Enzyme Formula** with every meal
- **Anti-inflammatory Enzyme Formula** three times per day
- **pH Balancing Formula** two times daily, morning and night

KIDNEY STRESS

The kidneys are the organs responsible for filtering the blood and excreting waste products and excess water in the form of urine from the body. The kidneys also regulate the pH, mineral ion concentration, and water composition of the blood.

The most important waste products for the kidneys to eliminate are those generated by the breakdown of proteins, which control the body's acid balance. These organs also regulate the production of red blood cells from the bone marrow, and convert vitamin D into an active hormonal form.

Kidney imbalances can result from genetic disorders, xenobiotics, including heavy metal toxicity, kidney stones, and bacterial infections.

Enzyme Supplementation Suggestions:

- **High Potency Digestive Enzyme Formula** with every meal
- **High Protease Enzyme Formula** three to four times daily

- **pH Balancing Formula** three times daily
- **Anti-inflammatory Enzyme Formula** may be added for inflammation or pain

LACTOSE INTOLERANCE

A deficiency in lactase, the enzyme responsible for digesting the lactose present in diary products, is common throughout the world. Approximately 70-90% of Black, Asian, American Indian, and Mediterranean adults are deficient in this enzyme. Many children lose their lactase by three to seven years of age.

Symptoms of lactose intolerance include flatulence, cramps, belching, a bloated feeling, and watery, explosive diarrhea.

Recommendations include avoiding or limiting the intake of dairy products, taking digestive enzymes with meals, and supplementing with probiotics to promote recolonization of the GI tract.

Enzyme Supplementation Suggestions:

- **High Lactase Formula** or **High Potency Digestive Enzyme Formula** with every meal
- **Probiotic Formula** daily

LARYNGITIS

Laryngitis is an inflammation of the mucous membranes of the larynx and irritation of the vocal chords. Laryngitis is characterized by dryness and soreness of the throat, hoarseness, cough and impaired speech. Often caused by a virus or bacteria, it may also be aggravated by smoking, allergic reactions, reflux, and straining of the voice.

Recommendations include ascertaining the underlying cause of the laryngitis, increasing hydration, rest, and supporting the immune system.

Enzyme Supplementation Suggestions:

High Potency Digestive Enzyme Formula with every meal
Soothing Digestive Formula open capsules, mix with water and drink
Anti-inflammatory Enzyme Formula may be taken anytime for inflammation

LIVER TOXICITY

The liver is the largest gland in the body. Located immediately beneath the right side of the diaphragm, it weighs nearly four pounds. This gland has many vital functions and plays a major role in metabolism. The liver produces bile that aids in the digestion of fats, the production of proteins for blood plasma, and the production of cholesterol and special proteins that help to carry fats around the body. The liver also stores glycogen, while filtering the blood of drugs and poisonous substances that would otherwise accumulate in the bloodstream.

The liver is extremely resilient. Up to 75% of its cells can be destroyed or surgically removed before it ceases to function. The most common cause of liver disease in the USA is alcohol consumption. Hepatitis is also a common liver disease.

Enzyme Supplementation Suggestions::

- **High Potency Digestive Enzyme Formula** and/or **High Lipase Enzyme Formula** with every meal
- **High Lipase Enzyme Formula** three times daily between meals
- **High Protease Enzyme Formula** four times daily between meals
- **Anti-inflammatory Enzyme Formula** may be added for inflammation

LONGEVITY [See AGING]

Longevity is defined as the length of a person's life. Research has produced many different theories of the aging process. Two predominant aging theories include damage and programmed theories. Damage theories advocate that the aging process is a result of cumulative damage to genetic materials and cells. Programmed theories believe that old age is determined by a genetic clock-like mechanism in the body. Although there are no magic cures to arrest the aging process, there are measures that one can take to slow the process and reduce the risks of premature death.

Significant factors that may contribute to an individual's longevity include: genetics, access to health care, hygiene, diet, exercise and lifestyle.

Recommendations in aiding longevity include a healthy, well-balanced diet, stress management, rest, regular exercise, support of the digestive and immune systems, and enzyme supplementation.

Enzyme Supplementation Suggestions:

- **High Potency Digestive Enzyme Formula** with every meal
- **High Protease Enzyme Formula** three times per day
- **Nattokinase Formula** two times per day
- **pH Balancing Formula** two times per day morning and night

LUPUS

Lupus is a chronic, inflammatory, autoimmune disease in which the body's immune system attacks itself. The specific cause of this collagen disorder has not yet been ascertained, however many researchers believe that it is due to a virus.

There are two different forms of lupus. Discoid lupus erythematosus affects the skin. A butterfly-shaped rash forms over the cheeks and nose that flares up with sun exposure. Additional symptoms of this disfiguring skin disease include small, soft groups of lesions that appear on the skin, producing scarring.

The second form of lupus is systemic lupus erythematosus, affecting the organs, blood vessels and joints of the body. Women develop lupus ten times more often than men, and fifty percent of individuals afflicted with system lupus erythematosus develop nephritis.

Both types of lupus periodically flare up and then go into remission. Infections, childbirth, excessive stress, fatigue, certain chemicals and drugs can precipitate attacks. Serious cases of lupus can adversely affect the kidneys, heart, nervous system and brain resulting in psychosis, depression, seizures and amnesia.

Recommendations include avoiding prolonged exposure to sunlight, lifestyle modification to reduce stress, promoting sufficient rest and sleep, dietary changes and enzyme supplementation to support the kidneys (low fat and salt), and digestive system and immune system.

Enzyme Supplementation Suggestions:

- **High Potency Digestive Enzyme Formula**
- **Anti-inflammatory Enzyme Formula** three to four times daily
- **pH Balancing Formula** three times daily
- **High Protease Enzyme Formula** three to four times daily

LYME DISEASE

Lyme disease is the result of a deer tick bite and is caused by infection from Borrelia burgdorferi bacteria. Sometimes it is misdiagnosed as Multiple Sclerosis, Rheumatoid Arthritis, Fibromyalgia, Chronic Fatigue Syndrome or other (mainly autoimmune and neurological) diseases. Lyme disease is considered one of the fastest growing infectious diseases in the US.

The first sign/symptom of this condition is generally the

appearance of a rash and a red papule on the skin a few days following the tick bite. Acute (early) reactions can include flu-like symptoms, fatigue, malaise, a stiff neck, backache, headache, joint pain and swelling, sinus infection, heart palpitations, and nausea and vomiting. Chronic (late) symptoms may include muscle twitching, seizures, panic attacks, depression, hallucinations and adrenal disorders. Left undetected, spleen and lymph node enlargement, arthritis, brain damage and an irregular heart rhythm can develop.

Symptoms usually subside slowly over a period of two to three years. Although antibiotic therapies are generally implemented, no cure has yet been found. If Lyme disease is suspected, consult with a physician as soon as possible.

Enzyme Supplementation Suggestions:

- **High Potency Digestive Enzyme Formula** with every meal
- **Anti-inflammatory Enzyme Formula** three times daily or whenever needed
- **High Protease Enzyme Formula** three times daily between meals
- **pH Balancing Formula** three times daily

LYMPHATIC CONGESTION

All body tissues are bathed in a watery fluid derived from the bloodstream. Much of this fluid returns to the bloodstream through the walls of the capillaries, but the remainder is transported to the heart through the lymphatic system. Lymph nodes are filters that trap microorganisms and other foreign bodies in the lymph. The nodes contain many lymphocytes (white blood cells) that neutralize or destroy invading bacteria and viruses. If an infection is particularly virulent, the lymphatics may become inflamed.

Enzyme Supplementation Suggestions:

- **High Potency Digestive Enzyme Formula** with every meal
- **High Lipase Enzyme Formula** three times daily
- **High Protease Enzyme Formula** three times daily
- **Anti-inflammatory Enzyme Formula** as needed for inflammation

MALABSORPTION SYNDROME

This imbalance involves the impaired absorption of nutrients, vitamins and minerals by the lining of the small intestine. If the intestinal tract mucosa lining is impaired or there is some blockage to the flow of digested foods through the muscle walls, then absorption becomes a problem. Without the absorption of these nutrients, a domino effect can occur in the system and take many different forms of ill health. Therefore, cleansing and support of the intestinal tract and mucosal lining is imperative.

Enzyme Supplementation Suggestions:

- **High Potency Digestive Enzyme Formula** with every meal
- **High Cellulase / Candida Formula** three times daily
- **Soothing Digestive Formula** three times daily

(to rebuild mucosal lining)
- **High Protease Enzyme Formula** or **Anti-inflammatory Enzyme Formula** may be added as needed

MENOPAUSE

Menopause occurs as the ovaries stop producing estrogen, causing the reproductive system to gradually shut down. Technically, menopause refers to the cessation of menstruation (ovulation), as a natural part of a woman's normal aging process. It is a time in a woman's life when physical and psychological changes occur.

The average onset of menopause is 50.5 years of age. Hot flashes and night sweats can be symptoms of menopause and occur in approximately 70% of all menopausal women. Additional symptoms may include headaches, atrophic vaginitis, frequent urinary tract infections, forgetfulness, and cold hands and feet.

Psychological symptoms are often attributed to the menopause, but it is not clear whether these symptoms are caused by the lack of estrogen or are a reaction to the physical symptoms and the sleep disturbances caused by the night sweats. The most common symptoms are poor memory, poor concentration, tearfulness, anxiety, and loss of interest in sex.

Changes in metabolism (internal chemistry) also occur during the menopause, but may not cause symptoms until later. The bones lose calcium more rapidly, especially in the first two to five years of the period. Other metabolic effects include a possible rise in blood pressure, and an increase in fats in the blood.

A common remedy offered for some of the symptoms is

estrogen replacement therapy. Because estrogen is associated with side effects and long-term risks, as well as benefits, a woman and her doctor must weigh the benefits against the risks before deciding whether to use estrogen replacement therapy. Side effects of estrogen therapy may include nausea, breast discomfort, headaches, and mood swings just to name a few. Whether estrogen might increase the risk of breast cancer has long been a concern. For women who are at high risk, it is not recommended. Postmenopausal women who take estrogen without progesterone have an increased risk of endometrial cancer (cancer of the lining of the uterus). The risk of developing gallbladder disease is modestly increased during the first year of estrogen replacement therapy.

Premature menopause occurs before the age of 40. Possible causes include a genetic predisposition and autoimmune disorders, in which antibodies are produced that can damage a number of glands, including the ovaries. Smoking has also been known to cause menopause.

Artificial menopause results from medical intervention that reduces or stops hormone secretion by the ovaries. These interventions include surgery to remove the ovaries or reduce their blood supply, and chemotherapy or radiation therapy to the pelvis to treat cancer. Surgery to remove the uterus (hysterectomy) ends menstrual periods, but it does not affect hormone levels as long as the ovaries are intact, and therefore should not cause menopause.

Nutritional needs become an important focus for menopausal women. The changes in the body's metabolic capability, hormonal structure, digestive capabilities, and just the aging factor cannot be stopped, but can be addressed to support the optimal nutritional potential available to keep a woman at peak performance.

Recommendations include incorporating a well-balanced diet, stress management, rest, regular exercise, in addition to dietary, glandular/hormonal and immune system support.

Enzyme Supplementation Suggestions:

- **High Potency Digestive Enzyme Formula** with every meal
- **High Lipase Enzyme Formul**a three times daily
- **High Protease Enzyme Formula** three times daily
- **pH Balancing Formula** three times daily
- **Anti-inflammatory Enzyme Formula** may be added anytime for inflammation

MIGRAINE HEADACHE

This imbalance involves an intense and disabling vascular headache that often lasts for two hours to two days. Migraines are characterized by severe pain on one or both sides of the head. They generally involve a throbbing or a pounding sharp pain, and are often accompanied by visual disturbances and/or nausea and vomiting. Migraines are common, affecting 15-20 percent of men and 25-30 percent of women. In a common migraine, the pain of the headache develops slowly, sometimes mounting to a throbbing pain that is made worse by movement or noise.

Classical migraines are relatively rare. A slowly expanding area of blindness surrounded by a sparkling edge precedes these headaches; these symptoms involve up to one-half of the field of vision of each eye. The blindness may resolve after 20 minutes, which is often followed by a severe one-sided headache with nausea, vomiting, hypersensitivity to light and hypersensitivity to sound. Other temporary neurological symptoms, such as weakness on one side, may also occur.

Migraine sufferers usually develop their own coping mechanisms for intractable pain. A cold or hot shower directed at the head, a wet washcloth, less often a warm bath, or resting in a dark and silent room may be as helpful as medication for many patients, but both should be used when needed. A simple treatment that has been effective for some individuals, is to place spoonfuls of ice cream on the soft palate at the back of the mouth. Hold them there with your tongue until they melt. This directs cooling to the hypothalamus, which is suspected to be involved with the migraine feedback cycle, and for some it can stop even a severe headache very quickly.

Causes of migraines include but are not limited to vascular instability, platelet disorders, nerve disorders, seratonin deficiency, certain drugs and xenobiotics, nutrient deficiencies, and allergenic foods.

Recommendations include ascertaining the underlying cause and/or precipitating factors of the migraines, dietary changes to identify and eliminate food allergies, stress management, adequate rest, and supplementation with enzymes and nutrients that may be deficient in the body.

Enzyme Supplementation Suggestions:

- **High Potency Digestive Enzyme Formula** with every meal
- **High Amylase / Allergy Formula** anytime symptoms arise
- **Anti-inflammatory Enzyme Formula** anytime symptoms arise
- **Nattokinase Formula** or **High Protease**

Enzyme Formula could also be substituted for the Anti-inflammatory formula

MUCOUS CONGESTION

The mucous membranes in the body are involved in absorption and secretion. They line the various body cavities and internal organs (such as the respiratory tract, gastrointestinal tract and genitals), in addition to other areas of the body that are exposed to the external environment (such as the nostrils, eyes, lips, ears and anus). They secrete mucus to maintain moisture and lubrication. Occasionally, certain substances causing inflammation and congestion can irritate these membranes.

Recommendations include ascertaining the precipitating factors or causes of the congestion, avoiding mucus-producing foods such as dairy products, use of drainage products to facilitate proper drainage of the congestion, the addition of lymphatic drainage therapies, increasing hydration and rest, and supplementation with enzymes and nutrients to support the immune system.

Enzyme Supplementation Suggestions:

- **High Potency Digestive Enzyme Formula** or **High Lactase Formula** (to ensure dairy breakdown) with meals
- **High Amylase / Allergy Formul**a three times daily (if the mucus is clear)
- **Seaprose / Mucolase Enzyme Formula** three times daily (if the mucus is colored)

- **High Protease Enzyme Formula** three times daily (to support the immune system)

MULTIPLE SCLEROSIS

MS is a chronic disease that affects the brain and spinal cord. This progressive, degenerative syndrome involves the gradual loss of the myelin sheath that surrounds the nerve cells (neurons) in the body. The neurons are the cells of the brain and spinal cord that carry information that create thought and perception, and allow the brain to control the processes of the body.

Multiple sclerosis can cause a variety of symptoms including changes in sensation, visual problems, muscle weakness, depression, difficulties with coordination and speech. Although many patients lead full and rewarding lives, MS can cause impaired mobility and disability in more severe cases.

Symptoms of this autoimmune disease may also include dizziness, sudden and transient sensory and motor disturbances, clumsiness, the "pins and needles" sensation, nausea, a feeling of spinning, loss of bladder sensation, and loss of sexual function.

Primarily affecting adults, most cases of MS begin between 20-40 years of age. Although there is no conclusive cause of MS, the following factors have been demonstrated to contribute to this disease: viral infection (i.e., measles), autoimmune reactions, a reduced capacity to detoxify free radicals, excessive lipid peroxidation, an excessive intake of saturated fatty acids and animal fats, and heavy metal toxicity.

Enzyme Supplementation Suggestions:

- **High Potency Digestive Enzyme Formula** with every meal
- **High Lipase Enzyme Formula** three times daily
- **High Protease Enzyme Formula** three times daily
- **pH Balancing Formula** three times daily
- **Anti-inflammatory Enzyme Formula** anytime needed for inflammation or pain

NAUSEA

Nausea is not an illness in itself; it is a possible symptom of several conditions, many of which are not related to the stomach at all. In fact, more often than not, nausea indicates a condition somewhere else in the body rather than in the stomach itself. Occasionally the stomach becomes upset, creating the sensation of a need to vomit. Additional symptoms may include excessive salivation, pallor and sweating.

Nausea can be caused by a variety of factors including reaction to certain foods, gall bladder imbalances, pregnancy, overeating, a reaction to certain medications, drugs, or therapies, in addition to a heightened sensitivity to motion, such as riding in an automobile, boat or airplane.

Enzyme Supplementation Suggestions:

- **High Potency Digestive Enzyme Formula** with every meal
- **Soothing Digestive Formula** after meals or any time it is needed

NERVOUSNESS [See STRESS or ANXIETY]

OSTEOPOROSIS

The major cause of osteoporosis is the gradual loss of protein matrix tissue from the bone. This loss results in pain in the back and hip, loss of height, increased fractures, and spinal curvature. The underlying mechanism in all cases of osteoporosis is an imbalance between bone resorption and bone formation. Either bone resorption is excessive, or bone formation is diminished.

Due to its hormonal component, more women suffer from osteoporosis than men, because their ovaries are no longer producing estrogen, which helps to maintain bone mass. Osteoporosis is also more common in smokers and drinkers and is associated with chronic obstructive lung disorders, such as emphysema and bronchitis.

Additional causative factors include lack of exercise, a calcium-phosphorus imbalance, lactose intolerance, and a diminished ability to absorb calcium through the intestines. Approximately 15 to 20 million Americans suffer from this disorder. While treatment modalities are becoming available, prevention is still the most important way to reduce fractures.

Enzyme Supplementation Suggestions:

- **High Potency Digestive Enzyme Formula** or **High Lactase Formul**a (if lactose intolerant)
- **pH Balancing Formula** three times daily
- **High Lipase Enzyme Formula** three times daily
- **High Protease Enzyme Formula** three per day

- **Anti-inflammatory Enzyme Formula** anytime inflammation or pain occurs

OXIDATIVE STRESS

Oxidative stress is a medical term for damage to animal or plant cells (and thereby the organs and tissues composed of those cells) caused by reactive oxygen species. It is defined as an imbalance between pro-oxidants and antioxidants, with the former prevailing. Oxidative stress is thought to contribute to the aging process.

Compounds that prevent free-radical damage are known as antioxidants. The body has antioxidant enzymes that include catalase, glutathione peroxidase and SOD (super oxide dismutase). These enzymes prevent damage caused by certain types of free radicals.

Symptoms of free radical damage (necessitating dietary antioxidant therapies) can include but are not limited to headaches, frequent or chronic infections, palpitations, neck spasms, back pain, or inability to relax, chronic allergies, diarrhea, nausea, heavy metal toxicity, chronic yeast infections or Candidiasis.

Laboratory studies are demonstrating that ingesting higher levels of antioxidants can increase life expectancy. Research is indicating that increasing levels of these compounds, which include vitamin C & E, betacarotene, selenium, coenzyme Q10, flavonoids, and sulfur-containing amino acids (cystein and methionine), can reduce the risk of developing cancer, heart disease, arthritis, macular degeneration, and other age-related degenerative conditions.

Enzyme Supplementation Suggestions:

- **High Potency Digestive Enzyme Formula** with every meal
- **High Protease Enzyme Formula** three times daily
- **pH Balancing Formula** three times daily
- **Anti-inflammatory Enzyme Formula** anytime inflammation or pain occurs

PARASITES

Any organism that spends a significant portion of its life in or on the living tissue of a host organism, and which causes harm to the host without immediately killing it is a parasite. Parasites also commonly show highly specialized adaptations allowing them to exploit host resources. Some parasites cause minor symptoms while others can cause extensive disease.

There are two kinds of parasites: endoparasites (those that live within their hosts) which include fungi (such as ringworm), and protozoa, (such as Giardia lamblia) and ectoparasites (those that live on but not within their hosts) which include ticks, fleas and lice.

Viruses and disease-causing fungi, and bacteria are all considered parasites. The outer coating of all of these organisms is protein. After they have been targeted by the immune system as a foreign agent, then the proteases, along with the immune cells, begin to remove them.

Enzyme Supplementation Suggestions:

- **High Potency Digestive Enzyme Formula** with every meal

- **High Protease Enzyme Formula** three to four times daily on an empty stomach
- **pH Balancing Formula** three times daily
- **Probiotic Formula** daily

PERIODONTAL DISEASE

Periodontal disease is a bacterial infection of the gums in the mouth, which can include gingivitis, bleeding gums, and pyorrhea. The symptoms can include one or more of the following: infection, abscesses, pain and inflammation in the gums, halitosis (bad breath), and in severe cases, infection of the bone resulting in bone destruction. The symptoms develop as a result of poor nutrition, excess sugar, plaque buildup, improper teeth care, drugs, glandular disorders, smoking, fillings that fit improperly, and an excessive consumption of alcohol.

Gingivitis involves inflammation and infection of the periodontal membranes around the base of the teeth and erosion of the bone holding the teeth. Periodontitis is generally the result of untreated gingivitis. This ailment is often related to a deficiency of calcium, folic acid, niacin, bioflavinoids or vitamin C. Additional causes of periodontitis may include poor diet, chronic illness, blood disease, smoking, drugs, excessive alcohol, and improper brushing.

Recommendations include regular dental care, dietary and lifestyle changes, and enzymes and nutrients to support the digestive and immune systems. [See CANKER SORES for additional information.]

Enzyme Supplementation Suggestions:

- **High Potency Digestive Enzyme Formula** with every meal
- **High Amylase Digestion Formula** may be added to any high carbohydrate or sugar foods
- **High Protease Enzyme Formula** three to four times daily between meals

PITUITARY IMBALANCES

The pituitary gland is a pea-sized structure that hangs from the base of the brain. This gland helps control aspects of physical growth, blood pressure, pregnancy (including breast milk production), thyroid and adrenal gland function, and overall metabolism.

Any abnormality of this gland usually means that it produces either too much or too little of one or more hormones and is causing changes elsewhere in the body. The hormones that may suffer include: ACHT, which stimulates hormone production by the adrenal glands; anti-diuretic hormone, which acts on the kidneys to decrease water loss in the urine and thus reduces urine volume; TSH, which is the thyroid stimulating hormone; the growth hormone, prolactin, which stimulates female breast development; melanocyte, which helps control the function of male and female sex organs; and oxytocin, which stimulates contraction of the uterus during childbirth and milk release from the breasts.

Enzyme Supplementation Suggestions:

- **High Potency Digestive Enzyme Formula** with every meal

- **High Lipase Enzyme Formul**a three times daily
- **High Protease Enzyme Formula** three to four times daily
- **pH Balancing Formula** three times daily

PREMENSTRUAL SYNDROME

Premenstrual Syndrome is the combination of various physical and emotional symptoms that occur in women the week or two before menstruation. This syndrome affects over 90% of women of reproductive age during their lifetime. Hormonal changes that occur throughout the menstrual cycle clearly influence PMS, but nothing concrete has been able to be established as the cause.

PMS may cause mood swings, irritability, tension, depression, anxiety, and fatigue. Physical symptoms may include breast tenderness, fluid retention, headaches, backaches, cramping, bloating, and lower abdominal pain.

Recommendations include dietary changes, lifestyle changes, and stress reduction or stress management. Also recommended are adequate rest/sleep and hormonal/glandular support.

Enzyme Supplementation Suggestions:

- **High Potency Digestive Enzyme Formula** with every meal
- **High Lipase Enzyme Formula** three times daily
- **High Protease Enzyme Formula** three times daily
- **pH Balancing Formula** three times daily

PROSTATE DISORDERS

The prostate is the male sex gland positioned beneath the urinary bladder. This donut-shaped gland encircles the urethra. During ejaculation, the muscles in the prostate squeeze fluids into the urethral tract. Disorders of the prostate rarely occur before the age of 30.

Prostate problems can include prostatitis (acute or chronic inflammation of the prostate gland), benign prostatic hypertrophy, BPH (which results in an enlarged prostate that can narrow the urethra), and cancer.

Symptoms of prostatitis include fever, frequent urination accompanied by a burning sensation, blood or pus in the urine, urinary tract infection, and pain between the scrotum and the rectum. As prostatitis becomes more severe, urination can become difficult.

Symptoms of prostate enlargement can include frequent urination, especially during the night, pelvic pain, burning, impotence, and difficulty stopping urination.

Prostate cancer rarely occurs in men under 60 years of age. Symptoms are vague and can include difficulty in starting urination, blood in the urine, a burning sensation during urination, and increasing frequency of urination at night.

Recommendations include thorough prostate examinations every year for men over 40 years of age, dietary changes (including avoiding foods with a high fat content), regular exercise and increased water intake.

Enzyme Supplementation Suggestions:

- **High Potency Digestive Enzyme Formula**
- **High Lipase Enzyme Formula** three times daily
- **High Protease Enzyme Formula** three times daily
- **pH Balancing Formula** three times daily

RESPIRATORY AILMENTS
[See ASTHMA and BRONCHITIS]

RHEUMATISM [See ARTHRITIS]

SENILITY [See LONGEVITY]

SINUSITIS

The sinuses are mucous membrane-lined air-filled cavities located in the facial region. These include the frontal sinuses, the maxillary sinuses, two sinus cavities located between the nasal cavity and eye sockets and the collection of air spaces in the large, winged bone behind the nose that forms the central part of the base of the skull.

Sinusitis is an inflammation of the nasal sinuses that accompanies upper respiratory infections. This ailment is generally caused most frequently through bacterial infections, although may be caused by viral infections as well. Smoking, irritating smells or fumes, nasal injuries and growth sin the nose, may cause chronic sinusitis. Sinusitis can also be caused by allergic reactions to foods [See ALLERGIES], particularly dairy products and hay fever.

Symptoms can include but are not limited to nasal congestion, fever, facial pain, headache, earache, toothache, general malaise, cranial pressure, clear or green/yellow mucous discharge, and a loss of the sense of smell.

Recommendations include ascertaining the underlying cause of the sinusitis, ingesting of hot liquids to relieve congestion and promote mucus flow and avoiding mucus-producing foods such as dairy products.

Enzyme Supplementation Suggestions:

- **High Potency Digestive Enzyme Formula** with every meal
- **Seaprose / Mucolase Enzyme Formula** three times daily
- **High Amylase / Allergy Formula** may be added for allergy assistance

SKELETAL PROBLEMS
[See BONE FRACTURES and OSTEOPOROSIS]

SKIN ERUPTIONS
[See also ACNE, DERMATITIS, and ECZEMA]

The skin is the largest excretory organ of the body and one of the primary sources of detoxification. Stress, poor digestion and enzyme deficiencies can cause a variety of skin reactions and eruptions, including acne (fats), eczema (sugars), and psoriasis (proteins). Fats trapped in the connective tissue under the skin can become cellulite.

Recommendations include determining the underlying cause of the skin disorders, dietary changes, stress reduction/management, detoxification if necessary, and supplementation with enzymes and nutrients to support the digestive and immune system.

Enzyme Supplementation Suggestions:

- **High Potency Digestive Enzyme Formula** with every meal

Consider:
- **High Lipase Enzyme Formula** three times daily (fats)
- **High Amylase / Allergy Formula** three times daily (sugars)
- **High Protease Enzyme Formul**a three times daily (proteins)

SLEEP [See INSOMNIA]

SOFT TISSUE INJURY

Tissues are a collection of cells specialized to perform a particular function. Muscular tissue consists of cells that are specialized to contract; epithelial cells form the skin and mucous membranes that line the respiratory and other internal tracts; nerve tissue comprises the cells that conduct electro-chemical nerve impulses; and, connective tissue includes blood, adipose tissue (fat), and various fibrous and elastic tissues (tendons and cartilages) that hold the body together.

Soft tissue injuries include sprains, strains, subluxation, repetitive stress injury, and carpal tunnel syndrome.

Recommendations include protecting the injured area.

Enzyme Supplementation Suggestions:

- **High Potency Digestive Enzyme Formula** with every meal
- **pH Balancing Formula** before bed

ACUTE:
- **Anti-inflammatory Enzyme Formula** anytime pain or inflammation until healed
- **High Protease Enzyme Formula** three to four times daily until healed

CHRONIC:
- **Anti-inflammatory Enzyme Formula** three to four times daily
- **High Protease Enzyme Formula** three to four times daily

STRAINS [See SOFT TISSUE INJURY]

SPLEEN DISORDERS

The spleen is closely associated with the circulatory system, where it functions in the destruction and removal of worn-out red blood cells, helps to fight infections, and holds a reservoir of blood. It is located in the upper left part of the abdomen, behind the stomach, and just below the diaphragm.

During times of stress when the oxygen content of the blood must be increased, the spleen reacts and releases its stored red blood cells into the blood stream. Surgical removal of the spleen due to injury or disease may increase one's susceptibility to infections.

Enzyme Supplementation Suggestions:

- **High Potency Digestive Enzyme Formula** with every meal
- **High Protease Enzyme Formula** three to four times daily
- **pH Balancing Formula** before bed

STRESS

Stress is a medical term for a wide range of strong external stimuli, both physiological and psychological, that can cause a physiological response called the General Adaptation Syndrome. Stress can have a major impact on the physical functioning of the human body by raising the level of adrenalin and corticosterone.

Stress creates an alarm in the body, and the system must retaliate by releasing certain hormones, such as cortisol, into the system to control the stress. These hormonal responses create a cascade of responses, all of which are meant to calm down the entire situation. Long-term stress can be a contributing factor to heart disease, high blood pressure, and stroke.

If the body is responding to stress, then obviously other tasks within the system are either being stopped or slowed down. Therefore, symptoms such as indigestion, headache, gallbladder stress, and increases in heart rate and blood pressure may manifest. One way to control stress is to maintain a daily exercise routine, which is crucial to the health of the whole system.

Recommendations include daily exercise, relaxation programs that may include yoga or physical massage and managing and controlling stressful situations.

Enzyme Supplementation Suggestions:

- **High Potency Digestive Enzyme Formula** with every meal
- **pH Balancing Formula** before bed
- **High Protease Enzyme Formula** before bed

SUGAR INTOLERANCES
[See DIABETES and HYPOGLYCEMIA]

The intake of excessive quantities of sugar (refined carbohydrates) can lead to many health disorders and diseases. Sugar, whether in the form of monosaccharides or disaccharides, stimulate the pancreas to produce insulin to metabolize carbohydrates. The greater the amount of sugar that is ingested, the more the pancreas must work to produce additional insulin. Over stimulation of the pancreas can wear out this vital gland and over time, hypoglycemia, hyperglycemia, and diabetes can result.

Symptoms of pancreatic stress can include fatigue, confusion, restlessness, dizziness, headaches, weakness, and adrenal exhaustion.

SURGERY

The surgical process can create a great deal of physical, emotional and mental stress. Supporting the digestive and immune systems prior to and following surgical procedures can facilitate and expedite the body's healing. Due to the extensive qualities of many surgical procedures and the medications that may be administered both before and after the surgery, it is best to consult with your physician regarding the use of enzymes and supplemental nutrients to assure that they are not contraindicated in any way.

Enzyme Supplementation Suggestions:

1-2 WEEKS PRIOR TO AND UNTIL 24 HOURS BEFORE SURGERY:
- **High Potency Digestive Enzyme Formula** with every meal
- **High Protease Enzyme Formula** three times daily
- **pH Balancing Formula** three times daily

AFTER SURGERY (ONCE FOODS CAN BE TAKEN BY MOUTH):
- **High Potency Digestive Enzyme Formula** with every meal
- **High Protease Enzyme Formula** three times daily
- **pH Balancing Formula** three times daily
- **Nattokinase Formula** three times daily

SWELLING [See INFLAMMATION]

SYSTEMIC LUPUS ERYTHEMATOSUS [See LUPUS]

THYROID IMBALANCE [See HYPOTHYROIDISM]

TINNITUS

Tinnitus or "ringing ears" is a phenomenon of the nervous system connected to the ear, characterized by a perception of a ringing, beating or roaring wound without an external source. The sound may be a quiet background noise, or loud enough to drown out all other sounds.

Common causes of tinnitus include presbycusis, prolonged exposure to loud environmental noise, and such pathological conditions as inflammation and infection of the ear, otosclerosis, Meniere's Syndrome and labyrinthisis.

Systemic disorders associated with tinnitus include hypertension and other cardiovascular diseases, neurological disorders including head injury, and hyper- and hypo-thyroidism. Tinnitus may also be exaggerated by severe emotional anxiety or physical stress.

Some cases will resist all conventional modes of therapy. Since this problem is so widespread, an association devoted to the study and management of tinnitus has been established. The address is American Tinnitus Association, P.O. Box 5, Portland, OR 9707.

Enzyme Supplementation Suggestions:

- **High Potency Digestive Enzyme Formula** with every meal
- **High Lipase Enzyme Formula** three times daily
- **High Protease Enzyme Formula** three times daily

TRIGLYCERIDES (Elevated)

Elevated blood triglycerides are now considered as important as high cholesterol in the development of ischemic heart disease. Triglycerides are compounds consisting of three molecules of fatty acids bound with one molecule of glycerol. Triglycerides play an important role in metabolism as energy sources. They contain more than twice as much energy as carbohydrates and proteins.

In the intestines, triglycerides are split into glycerol and fatty acids, and with the help of lipases and bile secretions, they move into the blood vessels where they are rebuilt in the blood from their fragments and become constituents of lipoproteins, which deliver the fatty acids to and from fat cells among other functions. Various tissues can release the free fatty acids and take them up as a source of energy. Fat cells synthesize and store triglycerides.

There are medium chain triglycerides (MCT) that range in length from 6- carbon chains and long chain triglycerides (LCT) that range in length from 8-4 carbon chains. The body uses these two types of triglycerides very differently. Research has indicated that MCT's do not promote weight gain and rapidly burn energy, promoting both weight loss and the burning of LCT's. LCT's on the other hand, are the most abundant fats in nature. These compounds are usually stored in the fat deposits of the body because they are difficult for the body to metabolize. The normal ranges for serum triglycerides are (60-160) for men and 35-135 for women).

Recommendations include dietary changes to lower fat intake and add regular exercise.

Supplementation Suggestions:

- **High Potency Digestive Enzyme Formula** or **High Lipase Enzyme Formula** with every meal
- **High Lipase Enzyme Formula** three times daily

- **Nattokinase Formula** three times daily

ULCERS (PEPTIC: Gastric, Duodenal, Esophageal)

A peptic ulcer is a sore on the mucous membrane of the gastrointestinal tract occurring in the stomach, duodenum, esophagus, or in the small or large intestine. [See COLITIS]

Ulcers are characterized by a loss of integrity of the area; secondary infection by bacteria, fungus or virus; a generalized weakness of the patient; and a lengthy healing time.

Symptoms of peptic ulcers can include abdominal tenderness and abdominal distress within an hour after meals or during the night. This distress is often relieved by ingesting food, taking antacids, or by vomiting. Other symptoms may include vomiting of blood, weight loss, or foul-smelling feces. Factors that can cause peptic ulcers include stress, smoking, poor diet, and food allergies.

Enzyme Supplementation Suggestions:

- **High Amylase Digestion Formula** with every meal
- **Soothing Digestive Formula** after every meal or taken as needed
- **pH Balancing Formula** three times daily

VAGINITIS [See also CANDIDIASIS]

This imbalance involves an inflammation and infection of the vaginal tract. Causes of vaginitis include Candida albicans (a yeast), often a result of antibiotic usage, certain types of birth control, certain microorganisms that are sexually transmitted including Trichomonas vaginalis (a protozoan), Gardnerella vaginalis (a bacterium), herpes simplex, Chlamydia trachomatis and Neisseria gonorrhea.

Symptoms of vaginitis include vaginal and vulval itching, irritation or burning, an increase in vaginal secretions, and secretions that have an abnormal color, odor or consistency, inflamed vaginal mucosa, and painful urination or pain after sexual intercourse.

Enzyme Supplementation Suggestions:

- **High Potency Digestive Enzyme Formula** with every meal
- **High Cellulase / Candida Formula** three times daily
- **High Protease Enzyme Formula** three times daily in chronic cases
- **Probiotic Formula** daily
- **pH Balancing Formula** three times daily may prove helpful

VIRAL INFECTIONS [See COLDS and FLU]

Viruses are one of the most frequent causes of infections. Because of their minute, microscopic sizes, these organisms can pass through even the tiniest cellular filters of the body. Once they enter the body, these opportunistic organisms make use of the existing cell host's enzymes and molecules to create more virus particles.

Symptoms of viruses include fever, muscle aching, chills and headaches. Examples of viral infections include the common cold, influenza, asthma, tonsillitis, measles, certain bladder infections, smallpox, encephalitis, croup, polio, AIDS, cold sores (herpes simplex), and mononucleosis.

Enzyme Supplementation Suggestions:

- **High Potency Digestive Enzyme Formula** with every meal
- **High Protease Enzyme Formula** three to four times daily
- **pH Balancing Formula** three times daily

WATER RETENTION [See EDEMA]

WEIGHT CONTROL

Weight control is a major concern for a large percentage of the American population. If you are overweight, it is wise to be concerned about your health and important to educate yourself regarding healthy and proper ways to lose weight. The true way to lose body fat requires time and effort. There are no miracle pills or easy methods.

The health risks of being overweight include: Type 2 diabetes, heart disease and stroke, osteoarthritis, gallbladder disease, high blood pressure, sleep apnea, gout, pulmonary problems, and GERD.

Recommendations include eating a healthy well-balanced diet, drinking lots of filtered water between meals, exercising regularly and supporting your endocrine and immune systems. Enzymes can assist the proper metabolism of nutrients, particularly fats or lipids.

Enzyme Supplementation Suggestions:

- **High Potency Digestive Enzyme Formula** with every meal
- **High Lipase Enzyme Formula** three times daily
- **High Protease Enzyme Formula** three times daily
- **pH Balancing Formula** three times daily

Appendix I
Enzyme Deficiency Test

The information presented in this questionnaire is intended to provide a profile of your past and present nutritional habits. It is not intended to diagnose, treat, cure, or prevent any disease.

Please circle the appropriate letter in each section.

SECTION 1
Which of the following best describes your body, especially when weight is gained?

A) Gain weight evenly (may then hold in stomach)
L) Carry weight in hips and thighs
P) Carry weight in upper body, especially the stomach
C) Has remained similar since teenage years (slim and trim, or heavy)

SECTION 2
In which category is your favorite food?

A) Carbohydrates (Vegetables / Breads / Pies / Sweets)

L) Rich Foods, Fatty Foods, Spicy Foods
P) Proteins (Meat)
C) Dairy

SECTION 3
Which foods give you problems?
Skip to the next question if food does not bother you.

A) Carbohydrates (Vegetables / Breads / Pies / Sweets)
L) Rich Foods, Fatty Foods, Spicy Foods
P) Proteins (Meat)
C) Dairy

SECTION 4
Please circle any issues you have had (present or past).

A) Allergies / Cold Hands and Feet / Depression / Fatigue / Headaches / Hemorrhoids / Low Blood Pressure / Neck and Shoulder Aches / PMS / Pancreatitis / Skin Eruptions / Sprue (Wheat Intolerance) / Upset Stomach / Ulcer

L) Aching Feet / Arthritis / Bladder Infection / Breast Lumps / Breast Tumors / Bypass Surgery / Cataracts / Cirrhosis / Cystitis / Eczema / Gallbladder Problems / Gallstones / Hay Fever / Hepatitis / Hives / Jaundice / Prostate Problems / Psoriasis / Urinary Problems

P) Alcohol Addiction / Arteriosclerosis / Back Problems / Candidiasis / Constipation / Ear Infections / Heart Disease / Herniated Disc / High Blood Pressure / Insomnia / Kidney Disease / Lower Back Ache / Loss of Hearing / Osteoporosis / Sciatica

C) Aching Knees / Chronic Allergies / Colds / Colitis / Crohn's Disease / Diarrhea or Constipation / Diverticulosis / Irritable Bowel / Milk Intolerance

ANSWERS

Which letter did you circle in each section?
Section 1＿＿＿＿＿＿＿＿ Section 2＿＿＿＿＿＿＿＿
Section 3＿＿＿＿＿＿＿＿

In section 4, determine under which letter you circled the most issues.
Section 4＿＿＿＿＿＿＿＿

DEFICIENCY TYPE

Two of any letter and one of another letter suggests you have both a dominant enzyme deficiency and a secondary enzyme deficiency (most common). Your secondary deficiency is the one in the section with the lowest number.

A pair of two letters suggests you have two enzyme deficiencies.

Three or four of any one letter suggests you have a dominant enzyme deficiency.

A different letter in each section suggests you are a type C (least common).

KEY

Type A is Amylase Deficiency (most common)
Type L is Lipase Deficiency
Type P is Protease Deficiency
Type C is Amylase, Lipase and Protease deficiency (Combination deficiency)

(Please note: You cannot be both a type C and another type.)

Suggestions for Diet, Exercise and Enzyme Supplementation

For Dominant Enzyme Deficiencies

Type A: (Amylase deficient)

Diet: Amylase is the enzyme that breaks down carbohydrates. Reduce intake of simple carbohydrates (cakes, pies, breads, pastas). Increase lean protein. If vegetarian, eat high protein plants.

Exercise: Low impact aerobic, three times a week. Walking is excellent for this type.

Enzymes: High potency digestive enzyme or high amylase enzyme blend with meals (best to take one of each, especially with high carbohydrate meals). High amylase product between meals to address problems associated with amylase deficiency.

Type L: (Lipase deficient)

Diet: Lipase is the enzyme that breaks down fats. Reduce intake of fatty and deep fried foods. Eat complex carbohydrates (vegetables) and lean proteins. Supplementing with flax oil, fish oil or both will help. In addition, take a high lipase supplement with oils.

Exercise: High cardiovascular exercises (depending on age), three times a week. Speed walking and jogging are excellent for this type.

Enzymes: High potency digestive enzyme with meals, which should contain no less than 2,500 FCCFIP of lipase. For health issues related to lipase deficiency, a high lipase supplement three times a day on an empty stomach (no less than 2,000 FCCFIP).

Type P: (Protease deficient)

Diet: Protease is the enzyme that breaks down proteins. Reduce protein intake. Increase complex carbohydrates. When eating protein, small, lean portions are best.

Exercise: Cross train (resistance and cardiovascular) at least three times a week for a minimum of thirty minutes each session.

Enzymes: High potency enzyme blend with meals (no less than 70,000 HUT). Add a high potency protease supplement with high protein meals, and also take between meals for maintenance.

Type C: (Combination deficient)

Diet: This is a deficiency in the enzymes that break down carbohydrates, fats and proteins. Moderation and balance with eating carbohydrates, fats, and proteins are important. Protein is best in the morning. Rotating food groups is beneficial for type C.

Exercise: Low impact aerobic and resistance training three times a week. Tai Chi would be a good choice.

Enzymes: A high potency digestive enzyme blend with meals and with snacks. A high protease supplement between meals for maintenance, adding a high amylase formula and high lipase formula when needed.

ADDITIONAL INFORMATION TO CONSIDER

For Secondary Deficiencies...

The recommendations above are for primary deficiencies. If you have a primary and a secondary deficiency, consider the above information and make the appropriate adjustments knowing you are deficient in more than one enzyme.

About enzymes with meals...

The number of capsules needed for each meal varies with circumstances. For example, a person in excellent health, with a strict diet that includes eating five times a day, may only need one capsule of a digestive enzyme blend with each meal. On the other hand, an individual in poor health (undergoing a severe health crisis), eating three large meals a day without doing any exercise, may need 2-3 capsules with meals. Use your good judgement when deciding how many enzymes to take with your meals, knowing you cannot take too many.

Appendix II
Protease Enzymes

Protease enzymes are the body's version of a Swiss army knife. They hydrolyze proteins, (long sequences of amino acids) into shorter chain peptides. This break down of protein into more manageable amino acids is the foundation of metabolism, digestion and all of life's processes. There are hundreds, if not thousands of different types of protease all having similar abilities. Some of the commonalities of protease are their ability to hydrolyze protein, have anti-inflammatory effects, increase circulation, support the immune system, dissolve blood clots, speed recovery, and decrease pain. Further, some individual protease enzymes are more outstanding at certain abilities than others. Consider just a few:

Nattokinase

Dr. Hiroyuki Sumi, M.D. a researcher of the Japan Ministry of Education was searching for a natural agent that could successfully dissolve blood clots associated with heart attacks and stroke. He found that natto exhibited a strong fibrinolytic (blood clot busting) activity. He named the corresponding fibrinolytic enzyme nattokinase. Dr. Sumi commented that nattokinase showed "a potency matched by no other enzyme."

Nattokinase is a novel fibrinolytic (fibrin dissolving) enzyme that is considered to be a promising agent for thrombosis therapy.[1] Nattokinase is derived from natto, a fermented Japanese soybean and has been the subject of numerous studies of safe use and effectiveness in dissolving clots and disorders associated with blood clotting. Fibrin is a protein that normally forms in the blood after trauma or injury to form a clot; however, it is also present when no such injury exists. This can cause high blood pressure along with circulatory and cardiovascular problems.

There are more than 20 enzymes in the body that assist in blood clotting, while only one enzyme, plasmin that can break a clot down. Nattokinase cleaves directly cross-linked fibrin in vitro.[2] Findings suggest that natto extracts, because of their thrombolytic activity, and suppress thickening after vascular injury as a result of the inhibition of mural thrombi formation[4] literally reducing the thickness of arteries. Further results indicate that the thrombolytic activity of nattokinase is stronger than that of plasmin or elastase in vivo.[2] Clinical trial suggests that nattokinase represents a possible option for use not only in the treatment of embolism but also in the prevention of the disease, since NK has a proven safety and can be mass produced.[3] Nattokinase also shows direst antioxidant qualities by having an inhibitory effect on the oxidation of low-density lipoproteins (LDL).[3] It is suggested that natto fractions might help to prevent arteriosclerosis, as they appear to reduce lipid peroxidation and improve lipid metabolism[4]. In conclusion, we've seen that nattokinase drastically reduces blood viscosity, which may have positive benefits for all those suffering from arteriosclerosis and atherosclerosis as well as hypertension, vascular disease and congestive heart failure. With its extremely safe and effective use, nattokinase may be the enzyme of choice for all those suffering from cardiovascular disease.

References:

1 Bioprocess Biosyst Eng. 2005 Dec 8:1-7 Purification of nattokinase by reverse micelles extraction from fermentation broth: effect of temperature and phase volume ratio. College of Bioscience and Bioengineering, Hebei University of Science and Technology, 050018, Shijiazhuang, Hebei Province, China.

2 *Biol Pharm Bull.* 1995 Oct;18(10):1387-91. Thrombolytic effect of nattokinase on a chemically induced thrombosis model in rat. Biotechnology Research Laboratories, JCR Pharmaceuticals Co., Ltd., Kobe, Japan.

3 *Acta Haematol.* 1990;84(3):139-43 Enhancement of the fibrinolytic activity in plasma by oral administration of nattokinase. Department of Physiology, Miyazaki Medical College, Japan.

4 *Experientia.* 1987 Oct 5;43(10):1110-1 novel fibrinolytic enzyme (nattokinase) in the vegetable cheese natto; a typical and popular soybean food in the Japanese diet. H. Department of Physiology, Miyazaki Medical College, Japan.

Seaprose

Seaprose is a semi-alkaline protease extracted from Aspergillus melleus. It is sold in a crystalline form, which means that it is super concentrated. The same enzyme is also available in a non-crystalline form, mucolase™. Both are measured in milligrams. It has all the wonderful abilities of the proteolytic enzyme family but stands out in its muco-lytic ability, literally meaning decreasing the symptoms of mucus, and shows potent anti-inflammatory qualities.

"There is clinical evidence that seaprose reduces sputum

viscoelastic properties in chronic hypersecretory bronchitis."[4] Seaprose's mucolytic ability stems from its action on muco-proteins and DNA, which are the major components of mucus that contribute to its viscosity, or stickiness. Seaprose shortens the "long chain of muco-proteins, DNA and other macromolecules, and thus reduces the viscosity of mucus facilitating its expectoration,"[1] thus improving the clearance of mucus from the bronchial tree. Furthermore the product does not seem to affect mucus glycoprotein secretion or secretory-IgA production, meaning it does not hinder the production of beneficial mucus.[4]

"In bronchial pulmonary infections antibiotics can be combined with other drugs, called mucoactive drugs, that act to reduce the abnormal viscoelasticity of the mucus enabling deeper penetration of more antibiotic into the mucus."[2] Seaprose is mucoactive and has a synergism between itself and other antibiotics allowing those antibiotics to become more effective in "performing a sterilizing action with therapeutic advantage."[2] "Results indicate that oral administration of seaprose stimulates the appearance of antibiotics in normal and inflammatory respiratory tracts and alveoli."[9] Seaprose "significantly increased the concentrations of these antibiotics in the bronchial wash."[9] Seaprose further shows a "post-mucolytic effect up to 8 days after treatment."[3]

Seaprose is not only endowed with proteolytic effect but its "anti-inflammatory activity has been tested in different clinical trials."[4] "In Freund's adjuvant-induced arthritis, seaprose-S significantly reduced the primary and secondary lesions."[5] Seaprose has been used in the treatment of bronchial inflammation, inflammation of ENT (ear, nose and throat) relevance, treatment of complications of puerperal surgical wounds (after childbirth), venous inflammatory disease."[4, 7, 8, 6]

Seaprose has also been studied in treatment with regard to safety. Seaprose causes "no adverse reactions," "is effective," "well tolerated" in trial, has shown "efficacy comparable to NSAID drugs but with greater safety."[4, 5, 6, 7]

References

1 In vitro Rheological Assessment of mucolytic activity induced by Seaprose, P.C. Braga, C. Rampoldi, A. Ornaghi, G. Cminiti, G Beghi and L. Allegra; 1990.

2 The influence of seaprose on erythromycin penetration into bronchial mucus in bronchopulmonary infections. Braga, PC; Piatti, G; Grasselli, G; Casali, W; Beghi, G; Allegra, L, 1992.

3 Effects of seaprose on the rheology of bronchial mucus in patients with chronic bronchitis. A double-blind study vs. placebo. Braga PC, Moretti M, Piacenza A, Montoli, CC, Guffanti EE, 1996.

4 Effects of seaprose on sputum [biochemical components in chronic bronchitic patients: A double-blind study vs. placebo. Moretti, M; Bertoli, E.; Bulgarelli, S.; Testoni, C.; Guffanti, E. E.; Marchioni, C. F.; Braga, P. C., 1995.

5 Anti-inflammatory effects of seaprose-S on various inflammation models. Fossati A. 1999.

6 Clinical study of the efficacy of and tolerance to seaprose S in inflammatory venous disease. Controlled study versus serratio-peptidase. Bracale G, Selvettella L., 1996.

7 The treatment of ENT phlogosis: seaprose S vs. nimesulide. Antonelli A, Cimino A, DiGirlamo A, Filippi P, Filippin S, Galetti G, Marchiori C, Marcucci L, Mira E, et al., 1993.

8 Clinical effectiveness and safety of Seaprose S in the treatment of complications of puerperal surgical wounds. Dindelli M, Potenaa MT, Candotti G, Frigerio L, Pifarotti G, 1990.

9 Effect of Oral Administration of a mold Protease on the concentration of antibiotics in rat bronchial wash. Junnosuke Kidaand and Kunio Kano. 1967.

Serratiopeptidase

Serratiopeptidase is a proteolytic enzyme (breaks down proteins) that has been shown to be absorbed into the bloodstream through the small intestine. Serratiopeptidase is also known as serratiapeptidase, serrapeptidase or serrapeptase.

This enzyme was originally found in the intestine of a Japanese silkworm. The silkworms use this enzyme to digest their cocoons. It has since been isolated from the gut of the silkworm and is now derived from the non-pathogenic bacteria known as Serratia E15, grown in much the same way Bacillus is grown to make nattokinase.

Serratiopeptidase is a multifunctional enzyme, helping with a variety of different disorders, especially pain and inflammation:

1) It helps to speed recovery time of the inflamed tissues by thinning fluids and helps increase drainage, which is formed by inflammation and injury.[1,2,3] It has also been shown to help block pain by inhibiting a specific type of amino acid called bradykinin.

2) In addition to being an anti-inflammatory it is also anti-edemic (prevents swelling and fluid retention). This enzyme also helps with cardiovascular health by dissolving fibrin, a sticky protein that forms in the blood. This fibrin leads to clots, blood coagulation and plaque in the arteries.[4]

3) Its application has also been used with respiratory conditions such as sinusitis. Serratiopeptidase efficacy with sinusitis has been recorded in many clin-

ical trials by thinning mucus secretions and helping with mucociliary transport and expectoration.[5]

4) In addition to all of the above actions, Serratiopeptidase can also help to digest non-living tissues such as a cyst or scar tissue, while posing no threat to any healthy living tissue.

In Germany and in other European countries, the use of Serratiopeptidase is standard treatment for arthritis, fibrocystic breast disease and carpal tunnel syndrome. Its use is slowly but surely being adopted by the United States. Studies are showing this enzyme to be a great replacement to any who use NSAIDs (Non-steroidal Anti-inflammatory Drugs) or those with respiratory conditions. It should be noted that in all the clinical trials on Serratiopeptidase there have been no known side effects; it has also been shown to be tolerated by individuals both young and old.

References:

1 A. Rothschild, J. Clinical use of serrapeptase: an alternative to Non-steroidal anti-inflammatory agents, *The American Chiropractor*. 1991, p. 17.

2 Esch PM, Gerngross H. Fabian A. Reduction of postoperative swelling. Objective measurement of swelling of the upper ankle joint in treatment with serrapeptase-a prospective study (German). *Fortschr. Med.* 1989. 107 (4), 67-8, 71-2.

3 Yamasaki H, Tsuji H, Saeki K. Anti-inflammatory action of a protease, TSP, produced by Serratia (in Japanese). *Nippon Yakurigaku Zasshi*. 1967. 63(4), 302-14.

4 Klein G, Kullich W. Short-term treatment of painful osteoarthritis of the knee with oral enzymes. A randomized, double-blind study versus diclofenac. *Clin. Drug Invest.* 2000. 19(1), 15-23.

5 Mohm. Maqbool, Professor and Head Akhtar Parvez, Registrar Sunhail Maqbool, Home Surgeon Dept. of E.N.T., Govt. Medical College, Srinagar-10.

Dipeptidyl Peptidase IV (DPP-IV)

Dipeptidyl peptidase (DPP-IV) is a protein that has multiple functions in the body. It is known as CD26 when it is on the surface of the T-cell (lymphocyte), which is known to support immune function. When found on the brush boarder mucosal membrane of the intestinal tract lining, it is known as DPP-IV. DPP-IV is a proteolytic enzyme that is able to break down an exorphin peptide (a protein) believed to be a contributing factor in gluten and casein intolerance. Though DPP-IV is technically a metabolic enzyme produced by the body, the activity has been discovered in a plant-based protease. It has been used nutritionally with success on individuals who have different levels of sensitivity to gluten and casein. Individuals with autism have been greatly helped by this enzyme.

Cause of Gluten Intolerance

For many individuals who consume wheat or dairy products, the particular proteins gluten and/or casein are difficult to digest and can lead to intestinal inflammation. In addition, a number of disorders including Celiac disease, Leaky Gut or other syndromes can result from intestinal inflammation.

When eaten, gluten exhibits a unique amino acid sequence which creates inflammation and a flattening of the villi of the intestinal tract. This becomes a serious disorder since the villi provide the intestine with the ability to absorb nutrients from the foods we eat. Damaged villi can lead to responses such as malabsorption, nutrient deficiencies or even disease.

The main treatment for gluten or casein intolerance has been to remove offending foods from a person's diet, also known as the Gluten Free, Casein Free (GFCF) diet. In the short term, this is effective because a person is removing the offending proteins, which will reduce inflammation and allow the villi to heal over time. But removing the food does not provide an ultimate solution. Once those foods are consumed again, even in cases of accidental consumption, the problems will repeatedly arise.

Formulas that contain DPP-IV along with other proteases offer those suffering with such intolerances an ability to literally digest and assimilate the offending proteins. This proactively heals the gut, reduces inflammation, allows proteins to be properly absorbed in their digested state, and broadens the amount of potential food groups a person may eat. DPP-IV may also be used by those who wish to remain on the GFCF diet to ensure complete break down of proteins safe and effectively.

From time to time a novel enzyme comes along that shows real promise. Most of the time it is a protease, although the following enzyme is actually an amylase.

Appendix III
Glucoreductase™ (GR)

Glucoreductase (GR) is an enzyme that is currently being studied for its ability to break down glucose. Enzymedica is researching the effects that this enzyme may have on people with type II diabetes. GR may reduce insulin needs in these individuals by reducing blood glucose. The theory is, that when GR is consumed with a meal, it can prevent carbohydrates from converting to glucose in the body. Since diabetes involves the need for insulin when glucose is present in the blood, it may reduce this need and assist individuals who are primarily controlling their diabetes through diet. Future research may prove that GR can assist insulin-dependent diabetics. Since the research is ongoing, it is difficult to say if and when this enzyme will be available in a formula.

Appendix IV
Protease and Probiotics - The Facts

The following represents a small number of the many available references that help answer the question: "Does protease kill probiotics?"

Enzymes (Including Protease) are Recommended with Probiotics by a Probiotic Expert

1. "There is a lot of discussion about whether probiotics can be taken with enzyme supplements. Let's think about that. We have discussed the use of enzymes that are naturally present in probiotics. In addition to lactase, probiotics contain lipases, *protease and peptidases* and many more enzymes. We know that probiotics are constantly breaking open and releasing their enzymes, which come in contact with other probiotics and human tissues. We also know that this causes no problem and, in fact is often a good thing. And we know that probiotics and enzymes are naturally present in fer-

mented foods and all eaten together. In sum, your probiotics and enzymes will function perfectly well together."

~ Dr. Mark A. Brudnak Ph.D, N.D. *The Probiotic Solution*, St. Paul, MN, Dragon Door Publications, 2003.

Protease is Recommended for the Treatment of Candida by Enzyme Experts

2. "Protease is effective in the prevention and treatment of allergies and Candida albicans"

    ~ Alison Heath, Food Enzymes – A Key to Unlocking Our Health *(M.D. News Magazine)*.

3. "When we take protease between meals to consume the excess radical protein, we have done away with the cause of Candida. The formulation I recommend is at least 600,000 units of protease activity three to five times a day. In the practice of enzyme nutrition, health professionals sometimes give over 3 million units of protease in one day."

    ~ Dr. DicQie Fuller, Ph.D., D.Sc. Founder of The Transformation Enzyme Therapy Center and Author of *The Healing Power of Enzymes*. New York, NY, Forbes Custom Publishing, 1998.

4. "The overgrowth of *Candida* is usually controlled by digestive secretions, including pancreatic

enzymes, bile, and hydrochloric acid. A deficiency of any of these essential digestive ingredients can permit excessive growth and yeast fungus. Therefore, it is critical to restore the gastrointestinal tracts normal digestive balance and secretions by furnishing assistance to the body with supplementation of enzymes..." (The author then goes on to recommend proteolytic enzymes as a therapy)

~ Dr. Anthony Cichoke. *The Complete Book of Enzyme Therapy* Avery Publishing Group, 1999, 226-227.

Protease Helps Probiotics

5. "Intracellular peptidase (protease) activities combine to provide the cell with critical amino acids. The activities of the proteolytic enzymes are therefore essential for the growth of LAB (Lactic Acid Bacteria)..."

    ~ Camilla Christenccon, Henrik Bratt, Lesley J. Collins, Tim Coolbear, Ross Holland, Mark W. Lubbere, Paul W. O"Toole, and Julian R. Reid. Cloning and expression of an Oligopeptidase, PepO, with Novel Specificity from Lactobacillus rhamnosus HN001.

6. "The scientific literature has been largely concerned with the impact of colonic fermentation products affected by the intake of prebiotics and probiotics (Macfarlane and Cummings, 1999) in regard to

colon cancer protection and systemic immunity. Absorption of microbial protein–derived amino acids would require that microbial protein breakdown occurs at the ileum. There is evidence for high proteolytic activity in human ileal effluents due to small intestinal peptidases but also due to bacterial proteases."

~ Cornelia C. Metges. German Institute of Human Nutrition, D-14558 Bergholz-Rehbrücke, *German Contribution of Microbial Amino Acids to Amino Acid Homeostasis of the Host.*

Probiotics Produce Proteases and are Present in Probiotic Cultures

7. "Lactic acid bacteria (LAB) are components of starter cultures for a variety of fermented products. The ability to produce extracellular proteinases is a very important feature of LAB. These proteinases catalyse the initial steps in the hydrolysis of milk proteins, providing the cell with the amino acids that are essential for growth of LAB."

~ D. Fira, M. Kojic, A. Banina, I. Spasojevic, I. Strahinic and L. Topisirovic. *Characterization of cell envelope-associated proteinases of thermophilic lactobacilli.* Institute of Molecular Genetics and Genetic Engineering, Belgrade, Yugoslavia 418/6/00.

8. "The proteases of lactic acid bacteria are generally localized extracellularly and peptide products (4-8

amino acids) are taken up into the cell. Limited studies on *Lactobacillus* species indicate that proteolytic systems exist in these bacteria. In fact, highly proteolytic strains of *Lact. helveticus, Lact. paracasei subsp. paracasei, Lact. acidophilus, Lact. casei, Lact. buchneri* and *Lact. delbrueki subsp. bulgaricus* have been identified."

~ A.L. Winters, J.E. Cockburn, M.S. Dhanoa and R.J. Merry *Journal of Applied Microbiology*, Volume 89, Issue 3, p. 442, September 2000, doi:10.1046/j.1365-2672.2000.01133.x "Effects of lactic acid bacteria in inoculants on changes in amino acid composition during ensilage of sterile and non-sterile ryegrass."

9. "Bacterial enzymatic hydrolysis has been shown to enhance the bio-availability of protein and fat (Friend & Shashani, 1984). Bacterial protease can increase the production of free amino acids which can benefit the nutritional status of the host particularly if the host has a deficiency in endogenous protease production."

~ Barry R. Goldin. Dept. of Family Medicine and Community Health, Tufts University school of Medicine, Boston MA. Health Benefits of Probiotics.

Amylases, which Break Down Sugar, can Feed the Candida Cell

10. "Carbohydrates come in different forms, but they all share the same basic makeup. What they look

like is not really important, but for the sake of simplicity, lets just say they are all sugar. Most sugars must be broken down to or converted to glucose to be utilized effectively. This is true of body cells, bacterial cells and yeast cells. Almost all types of life love sugar! And the simpler it is, the less energy it takes to break it down, and the faster it can be eaten and used as food."

~ Dr. Mark A. Brudnak Ph.D, N.D., *The Probiotic Solution*, St. Paul, MN, Dragon Door Publications, 2003.

Plant based proteases DO NOT BREAK DOWN ALL PROTEINS. The proteins that these proteases break down have to fit into one of three categories. They have to be 1) part of a dead organism, 2) damaged, or 3) attached to antibodies. These are the only proteins that these proteases have the ability to break down.

11. "Such foreign cells as bacteria and viruses are recognized as being foreign, as so-called antigens. When antibodies attach themselves to an antigen, an immune complex is formed. Every immune complex transmits signals indicating that other body defenses should come and destroy it. The ever veracious macrophages (white blood cells) are ideally suited to obey this command. However, all this is not sufficient for our bodies, better safe than sorry. The immune complex therefore alarms another death squad called the complement system as

well...They are, and this is easily forgotten, nothing more or less than another sort of our famous protein-degrading enzymes."

~ D.A. Lopez, M.D., R.M. Williams, M.D., Ph.D, M. Miehlke, M.D. *Enzymes: The Fountain of Life.* Neville Press,1994.

12. "Several studies have indicated that oral proteases bound to macroglobulins hydrolyze (break down) immune complexes, proteinaceous debris, damaged proteins, and acute phase proteins in the blood stream. It is suggested that oral proteases may hydrolyze and remove extracellular proteins damaged by free radicals."

~ Dr. DicQie Fuller, Ph.D., D.Sc. Founder of The Transformation Enzyme Therapy Center and author of *The Healing Power of Enzymes.* New York, NY, Forbes Custom Publishing, 1998.

13. "After oxidative damage (e.g., induced with iron, ascorbate, and oxygen), the inactivated glutamine synthetase is selectively hydrolyzed...These results suggest that protease (So) participates in the hydrolysis of oxidatively damaged proteins..."

~ Lee YS, Park SC, Goldberg AL, Chung CH. Protease So from Escherichia coli preferentially degrades oxidatively damaged glutamine synthetase. J Biol Chem, (1988 May 15) 263(14):6643-6.

14. "Our results show that (the serine protease) HtrA

plays a role in defense against oxidative shock and support the hypothesis that HtrA participates in the degradation of oxidatively damaged proteins localized in the cell envelope, especially those associated with the membranes."

~ Skorko-Glonek J, Zurawa D, Kuczwara E, Wozniak M, Wypych Z, Lipinska B. The Escherichia coli heat shock protease HtrA participates in defense against oxidative stress. *Mol Gen Genet*, (1999 Sep) 262(2):342-50.

INDEX

Diamond, Harvey, 26
diarrhea, 63, 84, 86, 87, 88, 164-165
diastase, 55. *See also* maltase
diet, 74, 76
   "ideal __," 43-44
   "nearly ideal __," 44-45
dietary supplements,
   see supplements
Digest Gold®, 65-66
digestion, 11, 32, 79
   definition of, 103-104
   enzyme formulas to assist
     with, 120-123, 127, 129-130
   and enzyme therapy, 39-45
   process of, 12, 15
   of proteins, 77
   problems with, 73, 82
dipeptidyl peptidase IV, 254
diverticulosis / diverticulitis, 165-166
DU, 66-67
duodenum, 13
dyspepsia, *see* indigestion

E

E. coli, 81
ear infection, 166-167
echinacea, 100
eczema, 17, 62, 87, 167-168
edema, 168-169, 252
electrolyte(s), 14, 73
embolism, 248
endocrine glands, 169-170
endocrine support enzyme
   formula, 128
endurance, 170-171
energy, 7, 8, 12 16
   for digestion, 40-45, 53
   enzyme formula to support, 128
   systemic, freeing up of, 101
   treatment to enhance, 170-171

See also biologically active
enteric coating, *see* supplements
environmental sensitivity, 63
Enzymatic Therapy™, 113
Enzyme Nutrition (Howell), 18, 42
enzyme therapy, 31-38, 42
   history of, 19-29
   plant-based ___, 37-38 47-52
   and immune system, 93-101
enzyme(s):
   animal-based, 24, 31-38, 39, 50
   use with animals, 20
   anti-inflammatory effect of, 22
   blend(s), 69, 78
   deficiency, 23, 39, 59
   definition, 7
   delivery system for, 78
   denatured (inactive), 8, 31, 33, 100
   digestive, 11-14, 42-43
   food __, 14-15
   formulas, 119-130
     see also supplements
   fungal (or microbial), 20, 37-38
   heat/temperature and, 8
   imbalances, 17
   metabolic, 11, 16-18, 42-43,
     49, 50, 97
   plant-based, 24, 26, 31-8, 39, 47-52
   potency of, 36, 65-70
   "potential," 42-43
   product companies, 113-114
   proteolytic, 58, 106
   systemic / therapeutic, 47-52
     see also enzyme therapy
   timing of usage, 47-48, 51
   types of, 9-18
   *See also individual enzymes
     such as* protease, lipase, etc;
     supplements, and enzyme
     therapy